Portrait of a Moral Agent Teacher

With anecdotes, classroom dialogue, and interview passages, this descriptive and detailed volume deeply explores the range of teaching practices that one schoolteacher intentionally uses to impart morality to her students. Discussed as teaching morally and teaching morality, these practices serve to reconceptualize moral agency as a comprehensive approach to moral education. Independent of any particular ethical, theoretical, or philosophical orientation, moral agency, so defined, embraces a range of assumptions, beliefs, and pedagogies, as well as teachers' folk wisdom, professional experiences, personal instincts, and moral understandings. Readers will be inspired by this grade-four classroom, and motivated to transfer several of the teacher's practices to their own school and teaching contexts.

Gillian Rosenberg received her PhD in Education from the Ontario Institute for Studies in Education at the University of Toronto, Ontario, Canada.

Routledge Research in Teacher Education

The Routledge Research in Teacher Education series presents the latest research on Teacher Education and also provides a forum to discuss the latest practices and challenges in the field.

Portrait of a Moral Agent Teacher

Teaching Morally and Teaching Morality

Gillian R. Rosenberg

Routledge
Taylor & Francis Group

LONDON AND NEW YORK

First published 2015
by Routledge

2 Park Square, Milton Park, Abingdon, Oxon OX14 4RN
711 Third Avenue, New York, NY 10017, USA

*Routledge is an imprint of the Taylor & Francis Group,
an informa business*

First issued in paperback 2017

Library of Congress Cataloging-in-Publication Data

Rosenberg, Gillian.
  Portrait of a moral agent teacher : teaching morally and teaching morality /
Gillian Rosenberg.
    pages cm. — (Routledge research in teacher education)
  Includes bibliographical references and index.
  1. Teachers—Professional ethics.   2. Teaching—Moral and ethical aspects.
3. Moral education.   I. Title.
  LB1779.R556 2015
  174'.937—dc23
  2015003355

ISBN: 978-1-138-79374-3 (hbk)
ISBN: 978-1-138-08491-9 (pbk)

Typeset in Sabon
by Apex CoVantage, LLC

To my parents, Allen and Zita Gardner.

Contents

Tables and Figures

Tables

Figures

Tables and Figures

Tables

Figures

Foreword

Teaching as a Moral Craft (Tom, 1984), *The Moral Dimensions of Teaching* (Goodlad, Soder, & Sirotnik, 1990), *The Moral Life of Schools* (Jackson, Boostrom, & Hansen, 1993), *The Moral Base for Teacher Professionalism* (Sockett, 1993), *Exploring the Moral Heart of Teaching* (Hansen, 2001), *The Moral Work of Teaching and Teacher Education* (Sanger & Osguthorpe, 2013): These book titles represent some of the scholarly contributions to the literature over the past three decades that have recognized and documented the moral nature of elementary and secondary school teaching. They have led the way for what is now, at least in the literature, commonly accepted to be true—that teaching is indeed morally infused. However, the work in this field is not by any means abundant. And, despite their, and others', insistence both empirically and theoretically about teaching's moral core and the central role and influence of the schoolteacher, the nuances of their vital messages about the moral essence of teachers' daily practice are often obscured, overlooked, or at best taken for granted in teacher education programs, in professional development initiatives, and in the normative discourse in schools themselves.

This seeming indifference is both perplexing and disappointing, as the moral aspects of teaching, it may be argued, should provide the very foundation of teacher professionalism as expressed both subtly and explicitly by how teachers conduct their practice and by what they teach of a moral nature to students. These two overlapping components of their professional agency—the "how" and the "what"—resonate with what the research in the field has defined as "teaching morally" and "teaching morality" (Fenstermacher, Osguthorpe, & Sanger, 2009). One could also distinguish them as *moral teaching* and *moral teachings*, two concepts that weave together to define what at least some if not many teachers actually do, consciously or not.

The latter component, *moral teachings*, refers to instruction broadly encapsulated by the idea of moral education in all its many forms. While traditionally more recognized than the embedded qualities of *moral teaching*, its focus is usually nonetheless on the moral curriculum or on the students and their moral development as a presumed result of their exposure to the curriculum. The emphasis is rarely centred on the person of the teacher. Consequently, the professional practitioner as ideally a moral agent, responsible through intentions, actions, and reactions, for both *moral teaching* and *moral teachings* remains inherently underconceptualized and underappreciated.

And yet, as I know from my own work with student teachers and experienced teachers alike, once they are introduced to the idea that daily teaching practice and all the aspirations, frustrations, indecisions, decisions, and exhilarations that accompany it are essentially morally charged, they enthusiastically embrace it. As Strike similarly observed of his student teachers, while they "rarely provide arguments that identify moral principles or seek to appraise ends, they do recognize the importance of such considerations once they are pointed out. . . . (They) find moral ideas familiar even though they are not the kinds of considerations that come to their minds first and they often need help in finding the words to express them" (Strike, 1993, pp. 103–104).

So, how can practitioners and policy makers who more often than not establish the contexts for teachers' work, as well as teacher educators, be "helped" not only to recognize but also to prioritize the moral nature of teaching? What might help immensely are vivid research portrayals of real moral agent teachers in real classrooms that connect the dots, so to speak, by illustrating that what to some are abstract concepts of moral philosophy are in actuality the core values that should and often do underlie everyday teaching. Moral principles such as fairness, honesty, empathy, respect, and so on (as well as, regrettably, their violation) characterize the planned and the spontaneous, the mundane and the exceptional, and the heartbreaking and the heartwarming aspects of what teachers experience on a regular basis. Recognizing this is a major step towards enhancing teachers' judgement and awareness of the moral significance of their actions and decisions. They first must conceptualize teaching practice as inescapably moral before they can improve the morality of their own practice. However, such research portrayals that merge philosophical analysis and empirical inquiry are not common, and those studies that do achieve this integration such as Jackson and colleagues' The Moral Life of Schools Project (1993) and Richardson and Fenstermacher's The Manner in Teaching Project (2000, 2001) stand out and endure over time as groundbreaking contributions to both the scholarly landscape and professional communities.

Dr. Gillian Rosenberg's important book, *Portrait of a Moral Agent Teacher: Teaching Morally and Teaching Morality*, extends this contribution in new and compelling ways. As an outstanding exemplar of qualitative research, it is conceptually sophisticated, empirically rich, structurally engaging, and eloquently accessible, as well as being unique and distinctive in its focus, context, methodological approach, and results. Rosenberg investigated the area of moral education as it is embedded in the daily and normative work of teachers through the powerful methodological lens of micro-ethnography.

In Chapter One, Rosenberg explains, "Positioning myself as a conduit, I metaphorically open the classroom door of one such teacher [a moral agent teacher possessing ethical knowledge], and broadly explore the following question: *How does a teacher who prioritizes the moral education of students envision, enact, and reflect on that moral education?*" Not only does she open the classroom door, she flings it wide and invites us in to experience in the most meticulous of detail the grade-four classroom of Ms. Terry Kennedy and her students. The single-case focus on one teacher over the course of an entire school year with a minimum of two full days of observation each week in the classroom sets it apart from other studies in the field and generated a depth and breadth of data that are truly remarkable. Readers will come to know Terry, see her, hear her, and picture for themselves the classroom, the students, and the daily routines. Rosenberg does indeed paint for us a "portrait" with clarity and tone, colour and texture. Such intensive and extensive fieldwork characterizes this book as being a highly original and distinctive study in its own right. Its elaborate and rich descriptive results in response to the overall research question further establish it as a very important and unique evolution in moral education research as it directly overlaps with the moral dimensions of teaching.

In this respect, Rosenberg's conceptual framework adopts a range of theoretical orientations and perspectives in new, creative, and empirically defensible ways both to make distinct and to show the interconnectedness of the moral practices of the teacher and the moral instruction of students. The analysis is organized carefully around four strands, or what she refers to as "domains", of teaching: teachers' vision for teaching, learning, and classroom life; moral content to be taught and learned; pedagogical strategies for teaching morality; and methods for assessing students' social-moral development. It weaves in theoretical concepts drawn from moral philosophy, moral psychology, and applied ethics as they relate directly to classroom life. The framework provides the foundation for Rosenberg's elaborate description of seven empirically determined themes illustrative of teaching morally and teaching morality. The result is not only compelling in its capacity to show how seemingly non-reconcilable theoretical concepts complement each other in practice, but also profound in its implications for teaching and teacher education.

Rosenberg remarked in the first chapter that this book is an adapted version of her PhD dissertation—I should point out that it is her *award-winning* dissertation; it won the 2013 Leithwood Award: the Outstanding Thesis of the Year at the Ontario Institute for Studies in Education, the University of Toronto. The external examiner at her doctoral defence was Professor Richard Osguthorpe of Boise State University, one of the leading scholars in the field who contributed to our appreciation of the "teaching morally–teaching morality" conceptualization. In his appraisal, he wrote:

> The thesis provides a fascinating framework for analyzing the moral work of teaching. More importantly, it brings that framework to life by describing the subject's classroom as one that is saturated with moral meaning and moral education. Those who question the relevance of the moral work of teaching will have a much more difficult time making such a claim in light of the findings in this thesis. The findings are unique in their focus on a single case study, providing one of the richest descriptions of the moral nature of teaching to date. And the field notes are exquisite—I felt like I was right inside the classroom. The author is to be commended for her remarkable ability to capture the moral life of the subject's practice in a way that leaves no doubt as to the authenticity of the findings. . . . The synthesis of strategies was excellent—long overdue in our field (especially those that espouse packaged character education programs). And the description of the overlaps kept the categories meaningful. . . . The categories that emerged are enlightening, and the analysis/synthesis of the data is powerful. The depth of the responses from the subject are amazingly rich and provide a layer of detailed analysis that is found in few if any other studies. The work will make an important contribution to the field in moral education and moral development. (Osguthorpe, 2013, personal communication)

And, indeed it will. It marks a notable research achievement, and I expect that its quality, significance, and originality as well as its practical contribution to teaching and teacher education will ensure its success as an invaluable research resource throughout academic and professional communities internationally.

Portrait of a Moral Agent Teacher is not a book about character education. However, it is a book about character and education—the professional character of Terry Kennedy, a moral agent teacher, and the education of a moral nature that she provides and enables. The combination of the two becomes far more than moral instruction; it becomes a kind of moral inspiration. As Terry often says about herself and her students, "I'm trying to grow them up", and this pursuit is grounded in moral meaning. Rosenberg reports that before the fieldwork commenced, a tentative Terry also said, "I know that I am not the perfect moral agent. There will be things I miss doing, and missed opportunities, for sure". However, that is exactly what makes this research so authentic, so real, and so important. Terry is on one hand admittedly an exceptional teacher and yet on the other hand she is also a recognizable teacher, ordinary in the best sense of the term, like many other teachers for whom moral agency is a daily reality as well as an ongoing aspiration. Thanks to the superb research skills and profound conceptual and analytical capacities of Gillian Rosenberg, you are about to be drawn into Terry's professional world—may it be for you as it was for me intellectually illuminating and emotionally rewarding.

Elizabeth Campbell, PhD
Professor
University of Toronto
January 2015

REFERENCES

Fenstermacher, G. D., Osguthorpe, R. D., & Sanger, M. N. (2009). Teaching morally and teaching morality. *Teacher Education Quarterly, 36*(3), 7–19.

Goodlad, J. I., Soder, R., & Sirotnik, K. A. (1990). *The moral dimensions of teaching.* San Francisco: Jossey-Bass.

Hansen, D. T. (2001). *Exploring the moral heart of teaching: Towards a teacher's creed.* New York: Teachers College Press.

Jackson, P. W., Boostrom, R. E., & Hansen D. T. (1993). *The moral life of schools.* San Francisco: Jossey-Bass.

Osgthorpe, R. (2013). Personal communication, unpublished dissertation assessment, January 18, cited with permission.

Richardson, V., & Fenstermacher, G. D. (2000). The Manner in Teaching Project. Retrieved May 10, 2000, from www.personal.umich.edu/~gfenster.

Richardson, V., & Fenstermacher, G. D. (2001). Manner in teaching: The study in four parts. *The Journal of Curriculum Studies, 33*(6), 631–637.

Sanger, M. N., & Osguthorpe, R. D. (2013). *The moral work of teaching and teacher education: Preparing and supporting practitioners.* New York: Teachers College Press.

Sockett, H. (1993). *The moral base for teacher professionalism.* New York: Teachers College Press.

Strike, K. A. (1993). Teaching ethical reasoning using cases. In K. A. Strike & P. L. Ternasky (Eds.), *Ethics for professionals in education: Perspectives for preparation and practice* (pp. 102–116). New York: Teachers College Press.

Tom, A. R. (1984). *Teaching as a moral craft.* New York: Longman.

Preface

September 10, 2008

Today is the start of a new term at school. I arrive early to class and get settled. As the other students enter the room, I observe them. One man strides in full of purpose and sits opposite the professor's chair but does not look up. I decide that he is likely shy, rather than aloof. Two women continue a conversation they began in the hallway. They laugh nervously and find seats opposite each other in the horseshoe of tables. Most, however, quietly glance around, and smile unassumingly as we make eye contact. Just before the start of class, a large man enters the room. I am immediately drawn to his broad smile as he returns mine, and the candid enthusiasm of his expression. I am happy when he sits beside me. We exchange only passing comments, as the professor introduces the course for the next 90 minutes.

When our break finally arrives, I can no longer contain myself. I must know who and what this man teaches. His reply is simple: "I teach my grade-eights how to live a good life and be good people". My astonishment must be obvious, as he continues to explain. "Of course, I use geography, history, science, and math to do that, but I am mostly concerned that they learn compassion". Barely regaining my composure, I nearly shout, "You are a moral agent to your students!" Now his face registers astonishment at the unfamiliar label I have given him.

Despite having observed and interviewed several teachers who were considered moral agents, I had not, as yet, met a teacher who so authentically and intuitively identified with this role. Whether brought together by fate, luck, or mere coincidence, we realized in that brief exchange how much we would learn from each other.

As I read and write about moral education for children; as I parent my own children and communicate with their teachers; as I dialogue with teachers who are fellow students, friends, and relatives; as I listen to my children discuss what they like and dislike about their teachers; and as I spend time in classrooms, I am more and more aware that unsung heroes walk among us—teachers who make a positive impact on the people their students will become. Such teachers do not need to be instructed by others on the pro-social and moral values children should be taught to embody. They do not need to be told that they should enter into trusting and respectful relationships with their students. They do not need policy to infuse their practices and the learning environment with ethical values and moral principles, and they certainly do not need a formal program to show them how to foster students' social-moral growth and development. My vision for this work was born from these observations, experiences, and beliefs. I wanted to find such a teacher and create an opportunity for that teacher's practices to inspire and be instructive to others.

> Good teachers put snags in the river of children passing by, and over the years, they redirect hundreds of lives. Many people find it easy to imagine unseen webs of malevolent conspiracy in the world, and they are not always wrong. But there is also an innocence that conspires to hold humanity together, and it is made of people who can never fully know the good that they have done. (Kidder, 1989, p. 313)

REFERENCE

Kidder, T. (1989). *Among Schoolchildren.* Boston: Houghton Mifflin Company.

Acknowledgements

The research project that led to this portrait was so thoroughly enjoyable because of Terry Kennedy (pseudonym), my partner in generating understandings and insights. Terry is a genuine, humble, and soulful person, and a moral exemplar for adults as well as children. I am a better person for knowing her. I thank her for having courage to allow me into the classroom, and for the openness and honesty with which she revealed her practices and disclosed her beliefs and assumptions. I am also grateful to the 15 students who shared grade-four with me and embraced me as part of their class community.

Speaking of moral exemplars, I acknowledge with much appreciation the support of my mentor and friend, Dr. Elizabeth Campbell, who kindly provided a foreword for this book. Elizabeth's exceptional academic work is cited in every chapter. Her support and guidance, although more implicit, is ubiquitous. I also thank my friend Cameron Bond, whose intellect I co-opted for the onerous task of editing the first draft. I am grateful for the oversights he identified, from typos to repeated content to comments requiring more substantiation.

Finally, I am blessed with a family who provides much love and support. My three children, Madeleine, Charlie, and Benjamin, kept me company at coffee shops and libraries, doing their schoolwork while I did mine. Stories of their classroom experiences, which they shared with honesty and passion, helped to shape my general interest in the moral dimensions of schooling. My husband, David Rosenberg, is a writer by craft and trade. He taught me how to use commas and quotation marks sparingly, and that I *can* begin a sentence with the word *and* (or any word, for that matter) if it enables me to best express the sentiment. I have finally had the courage to do so in my academic work. See if you notice where. For reading everything I write, and always letting me be the person I want to be, I thank him most deeply.

1 Spending the Year
A Micro-Ethnographic Study

The essence of civilized life resides in a moral code that inhibits and restrains savagery and barbarism. Sizer and Sizer (1999) identify four means by which this code is transmitted, from one generation to the next: the family or clan, religion, media, and schooling. In the past, family, clan, and religion assumed primary responsibility for the moral education of children. Today, in Western societies, both mechanisms are considered private, and neither parents, other family members, nor religious leaders are held publicly accountable for the moral development of youth. In the past, media was limited in reach and restricted in content. Today, it is boundless and often replete with unethical and immoral messages (Ryan & Bohlin, 1999; Sizer & Sizer, 1999). Schooling, therefore, has become the focus for restoring a morally balanced influence on children and youth. Accordingly, schools are societally pressured and in many jurisdictions politically compelled to provide moral education (Arthur, 2008; Leming, 2001; Lickona, 1991; Ryan & Bohlin, 1999; Setran, 2003; Stengel & Tom, 2006).

Secular forms of moral education emerged, in earnest, by the 20th century. In the latter half of that century, educators, primarily in the United States, were exploring an assortment of approaches, with variable degrees of success (Balch, Saller, & Szolomicki, 1993; Joseph & Efron, 2005). Traditional character education, in particular, dominated the late 1980s and arguably remains dominant today. Teachers were attracted to its direct methods, which coincided with leading conceptions of teaching, learning, and class management, including teaching as knowledge transmission; learning as the passive acceptance of knowledge; and discipline as behaviour control (Ryan & Bohlin, 1999; Watson, 2008). Watson (2008) remarks that "[character education] did not require a rethinking of the whole educational endeavour. Whether transmitting values or math skills, the education process of telling, modelling, explaining, practice and correction would be the same" (p. 178). Further, character curricula could be externally developed, packaged, and provided to teachers in a format ready for delivery in the classroom. This facilitated consistency among teachers, schools, and school boards, without consuming an inordinate amount of their time and energy. Finally, there was a growing system of support for educators, including the emergence of four large character education organizations in the United States; the establishment of college and university centres dedicated to character education; the prevalence of related keynote conference addresses; and the publication of books, magazines, learning resources, and other relevant materials. Ryan and Bohlin (1999) observe, "What was once a modest movement has become a thriving industry" (p. xv).

Yet, despite these benefits, six White House congressional conferences on character education (Ryan & Bohlin, 1999), and political mandates to implement character programs in Canadian schools, scholars and education stakeholders questioned the theoretical foundations of character education (Davis, 2003; Kohn, 1997), criticized its narrow approach to moral education (Joseph & Efron, 2005), and challenged the lack of empirical evidence to support outcome claims (Berkowitz, Battistich, & Bier, 2008; Leming, 2008; Nucci & Turiel, 2009). Referencing an earlier incarnation of character education, Dewey (1959) observed:

> The influence of direct moral instruction, even at its very best, is *comparatively* [italics in original] small in amount and slight in influence, when the whole field of moral growth through education is taken into account . . . the development of character through all the agencies, instrumentalities, and materials of school life. (p. 4)

The field to which Dewey refers comprises the multiple moral dimensions of teaching, learning, and classroom life. Empirical studies have reported that some teachers knowingly, intentionally, and systematically embed moral values of respect, honesty, fairness, inclusiveness, kindness, compassion, sensitivity, integrity, trustworthiness, and courage, among others, in the everyday business of establishing rules, rituals, and routines; managing events and activities; disciplining students; creating and maintaining a psychologically safe classroom environment; forging relationships with and among students; attending to students' particular learning needs; implementing cooperative pedagogy; and instilling a sense of citizenship and social responsibility (Campbell, 2003, 2004; Campbell & Thiessen, 2000, 2001; Fallona, 2000; Richardson & Fenstermacher, 2000, 2001). Dewey (1959) succinctly captured the spirit of this, saying, "The teacher who operates in this faith will find every subject, every method of instruction, every incident of school life pregnant with moral possibility" (p. 58).

The ability to recognize and act upon such moral possibilities entails ethical knowledge, defined as teachers' awareness, appreciation, and application of one's personal moral wisdom to professional practice (Campbell, 2003). Addressing moral education in schools, therefore, seems to reside, first, in nurturing ethical knowledge among teachers, and only second, in character education programs or other formalized approaches to moral education. As Weissbourd (2003) states, "The moral development of students does not depend primarily on explicit character education efforts but on the maturity and ethical capacities of the adults with whom they interact" (p. 6). Teachers who possess ethical knowledge currently operate within schools and classrooms. Yet, the depth of their ethical knowledge and the complexities of related practices remain elusive, hidden behind closed doors in a professional culture that is traditionally private and personal. Positioning myself as a conduit, I metaphorically open the classroom door of one such teacher and broadly explore the following question: *How does a teacher who prioritizes the moral education of students envision, enact, and reflect on that moral education?* This book recounts what I have learned and what I believe to be informative and inspirational regarding school-based moral education.

The rest of Chapter One establishes foundations for appreciating the discussions that follow. Terms that have already been used and will continue to be used are clarified, and I directly address my position on moral agency, which is more fully developed in the final chapter as the portrait's core concept. The empirical study that generated this account is also outlined, introducing the teacher participant and identifying methods of data collection. I candidly share my personal beliefs, assumptions, and ethical perspective to help contextualize the study's results, which were intellectually and psychologically distilled by me. Finally, I provide a roadmap of sorts, briefly previewing how the topics and themes are organized in the chapters that follow.

DEFINITIONS

This book represents an empirical report, rather than a philosophical treatise. Accordingly, it is not grounded in moral philosophy or meta-ethics, but rather in applied, normative morality. I draw the following definitions, therefore, from the moral dimensions of teaching literature, as related to the moral nature of teaching practice, and from moral education literature, as related to how morality is conveyed to others, typically youth. Several philosophical meanings are referenced, nonetheless, as they support the construction of applied meanings.

Virtues, Morals, and Ethics

In the context of moral philosophy, virtues, morals, and ethics are carefully distinguished, as follows: *Virtues* are consistent traits and dispositions that define one's character in terms of goodness, righteousness, and excellence; *morals* are concerned with principles of right and wrong and good and bad, which guide one's attitudes, motivations, and behaviours; and *ethics* refers to systems or codes of conduct, the philosophical study of morals, and collective and professional interpretations of moral standards (Campbell, 2003; Carr, 2008). These philosophical distinctions are not overly relevant in the applied context of this work. As is common in professional literature and some philosophical literature, I use the terms interchangeably to represent what is right, good, caring, and virtuous in one's actions, character, thinking, relationships, and ways of being.

Often my choice of term is based on how comfortable and familiar a particular phrase may be. For example, the terms *moral education* and *moral agency* are more typically used than *virtues education* and *ethical agency*. Similarly, the terms *code of ethics* and *professional ethics* are preferred over *code of morals* and *professional virtues*. Yet, I tend to favour the term *moral*, overall, despite Sockett's (2012) observation:

> There has been a strong tendency in American culture to regard the word moral with disapproval, so much so that it is sometimes seen as inflammatory. Often the word "ethical" is used because of the bad press the word moral has encountered. (p. 160)

For me, the word *moral* cuts to the core of humanity and represents the essence of what enables human flourishing, welfare, well-being, and dignity. As such, it coincides more closely with how this work is situated in teachers' personal and professional moral orientations, rather than institutional codes, and in moral education, rather than teacher professionalism and professionalization. Additionally, I consider the terms *moral values* and *ethical values* to be synonymous with the terms *virtues*, *morals*, and *ethics*. I use them interchangeably, as well, to refer to a subset of values that are moral in nature.

Moral Education

Moral education, in the context of schooling, is identified as any opportunities teachers, administrators, and staff undertake to knowingly, purposefully, directly, and formally communicate lessons and messages of a moral nature; and any coincidental, inadvertent, indirect, and informal opportunities students have for learning about morality. The former is referred to as teaching *morality*, and the latter is association with teaching *morally*. This distinction was advanced by Fenstermacher, Osguthorpe, and Sanger (2009) and is fully developed in Chapters Four and Five by way of characterizing this portrait.

In education literature, the term *character education* is sometimes used synonymously with the term *moral education* (Howard, Berkowitz, & Schaeffer, 2004). At other times, they are distinguished by capitalizing the term *character education* when referring to a program rooted in normative virtue ethics and using lower case when referring, more generically, to moral education (Davis, 2003; Fenstermacher & Osguthorpe, 2000). I reserve the term *character education* for the former, as one orientation toward moral education. Further, although *moral education* is broadly defined in terms of teaching and learning, this book, and the study on which it is based, focuses on what is taught, rather than what is learned; on teachers, rather than administrators and staff; on school, rather than home and other institutions; and on the classroom, rather than the school environment. These additional topics have been explored by others and are significant to the present work, although they cannot be reasonably accommodated within its scope.

Moral Agency and Ethical Knowledge

Moral agency has roots in Kant's notions of rational autonomy and in Aristotle's notions of virtue ethics. Contemporary literature draws on both. Sockett (2012) states, "The moral agent is, formally speaking, someone who acts consciously in pursuit of valuable ends or according to rules of conduct" (p. 159). Fenstermacher and Osguthorpe (2000) note, "Agency is accomplished by means of deliberation on proper means and worthy ends. A moral agent is one whose choices can be accounted for by the giving of reasons, and these reasons explain and justify their choices" (p. 9). Boostrom (1998) claims that an agent is praiseworthy or blameworthy only if he or she had "more than one course of action available, as well as both the authority and the competence to choose which course of action to follow" (p. 181). Finally, Lickona and Davidson (2005) suggest that moral agent teachers sustain the following qualities (p. 147):

- Respect the rights and dignity of all persons;
- Understand that respect includes the right of conscience to disagree respectfully with others' beliefs or behaviours;
- Possess a strong sense of personal efficacy and responsibility to do what is right;
- Accept responsibility for mistakes;
- Accept responsibility for setting a good example and being a positive influence;
- Develop and exercise capacity for moral leadership.

Campbell (2003) further develops these definitions to create a model of moral agency as a *double-pronged state*, concerned with both "how teachers treat students generally and what they teach them of a moral and ethical nature" (p. 2). This unifies Fenstermacher's (2001) concepts of moral agency and moral development, which he defines as follows: "Moral agency is that quality possessed by a person to act morally. Moral development is the bringing about in others of moral agency" (p. 650). Despite this philosophical distinction, Fenstermacher acknowledges an applied connection between the two, noting that "there certainly seems to be some sort of requirement for moral agency on the teacher's part in order to be well and effectively engaged in the moral development of students" (p. 651). Campbell (2003) formalizes this requirement by suggesting that the moral agent teacher is both a moral person and a moral educator.

As a moral person, the teacher attends to ethical and moral standards regarding one's personal conduct in the classroom and at school, one's interactions with students and colleagues, and the carrying out of one's professional responsibilities. Embedded in this prong of moral agency is the assumption that professional practice is linked to personal morality, with the non-consequentialist argument that students have a right to experience an ethically conceived and ethically delivered education. Providing such education extends beyond the professional rules, codes of conduct, and moral expectations intrinsic to the role of teaching (Campbell, 2003). There is also a consequentialist aspect to this first prong of moral agency that anticipates the second prong—the teacher as a moral educator. Campbell (2003) states, "Teachers, through their actions, words, and attitudes, may be seen to be living by the same principles that they hope students will embrace" (p. 2). This identifies teachers as moral role models and exemplars. As moral educators, however, the teacher also intentionally provides moral instruction and moral direction to students, through, for example, spontaneous messages, formal lessons, disciplinary action, programs, and rules.

Both prongs of moral agency are underpinned by *ethical knowledge*. Campbell (2003) coined this term in reference to a type of professional knowledge sustained by teachers, which makes visible the conceptual and practical connections between teachers' personal interpretations of moral values and virtues, and their professional obligations, duties, roles, and responsibilities; and which brings to consciousness the moral complexities, conflicts, and nuances of classroom life, teaching practices, and learning goals. With this conception of ethical knowledge, moral agency is considered an intentional professional expectation and, as such, "more than an inevitable state of being, created by circumstances that bring adult teachers and children together in a learning environment" (Campbell, 2003, p. 2).

I derive my understanding of moral agency from Campbell's (2003) model, revised to reflect a different focal point. Campbell's interests reside in teacher professionalism and ethical practice. Although related, my interests reside in the moral education of children. Accordingly, I consider moral agency to be an active, conscious, and purposeful means by which teachers provide moral education, such that moral agent teachers hold, as a primary concern, the moral development of their students. This definition emphasizes the moral educator prong, while implying the moral person prong. Teachers who deliver a prepackaged character program, for example, are only considered moral agents if they also conduct their professional practices as moral people. Alternatively, teachers who conduct their professional practices as moral people may not prioritize or see as their role the facilitation of students' moral development, or they may routinely overlook moments for communicating moral messages and lessons. Such teachers do not display the moral education prong and, therefore, are not considered moral agents.

This definition assumes that a teacher is or is not a moral agent. As long as both prongs are represented in the teacher's professional practice, however, different degrees of moral agency might be recognized. Fensterm-acher (2001) suggests, "One's facility with the methods of moral develop-ment may be intimately connected to the depth and sophistication of one's moral agency" (p. 650). Accordingly, he speculates on "how morally good a person need be in order to be good at cultivating moral goodness in stu-dents" and "how fully developed a moral agent must be in order to be good at moral development" (p. 650). Positioning ethical knowledge as the knowledge base of moral agency substantiates this line of inquiry and makes tangible the possibility of nurturing moral agent teachers, a notion I return to in the final chapter.

BELIEFS, ASSUMPTIONS, AND ETHICAL PERSPECTIVE

I began my professional life in the quantitative world of laboratory genetics, as a technologist and later a college instructor. In these roles, I operated on moral autopilot, giving limited conscious thought to the moral dimensions of my practice, or my personal and professional conduct. Early graduate studies in education administration focused on leadership, as well as orga-nizational climate and culture. Although I was beginning to explore ethical underpinnings of both, the applications were primarily in higher educa-tion. Ultimately, I developed a keen interest in kindergarten to grade-12 schooling, children's experiences at school, and everything moral related to teaching, learning, and classroom life. I attribute this to three personal expe-riences, which I will recount briefly, as they consciously and unconsciously informed the beliefs, assumptions, and ethical perspective that are the lenses through which I filter and analyzed data, and present results and insights.

First, years of graduate studies in education afforded many opportuni-ties to connect with schoolteachers who enthusiastically shared morally salient anecdotes and engaged in discussions of a moral and ethical nature. Through their eyes, I began to understand and appreciate morality as foun-dational to the professional life, role, and responsibilities of teachers. Sec-ond, toward the beginning of a master's degree I participated as a research assistant on a qualitative project, called The Moral and Ethical Bases of Teachers' Interactions with Students (Campbell, 2003, 2004; Campbell & Thiessen, 2000, 2001). At the time, there was no obvious connection to my skills, knowledge, background, or interests, a point made embarrassingly evident when, at team meetings, I asked questions such as, "What is your hypothesis?"; "What are you trying to prove?"; "How are you going to measure the variables?"; and "What are the variables?" Eventually find-ing my feet, I was exhilarated by qualitative methods of investigation, and I loved being in the classroom with children. In addition, I was deeply moved by the efforts of participating teachers to attend to their students' personal needs and social-moral growth and development. Finally, my own children's schooling has exposed me to ethical and unethical conduct, behaviours, and practices among teachers; it has also exposed me to teachers whose awarenesses of the moral dimensions of teaching, learning, and classroom life range from enlightened to indifferent or ignorant. Accounts of teachers who have been unethical or of teachers who are morally indifferent or ignorant are cause for great concern. When in the class of such a teacher, my children expended much energy to contain feelings of frustration, unhappiness, anxiety, insecurity, powerlessness, and inadequacy, and to socially manage the negative fallout among their peers. Their learning potential and academic achievement was compromised, as was their sense of self and self-confidence, and emotional and social well-being. This was not the case, however, when in the class of teachers on the other end of this moral spectrum.

These three experiences present different viewpoints of the same phenomenon—the teacher's, the researcher's, and the student's. Subsequently, I have come to believe the following statements: All acts of teaching and learning and all aspects of classroom life have a moral core. This moral core can be identified, empirically studied, and made visible. Teachers can attend to the moral dimensions of teaching, learning, and classroom life, and in fact, should do so to properly achieve desired learning outcomes and to provide students with an educational experience that is morally justifiable. Finally, these moral dimensions directly and indirectly, and favourably or unfavourably impact students in a personal way. For me, these beliefs are intellectual, psychological, and emotional. They awakened a passion within, ultimately redirecting my academic and professional interests. By sharing this personal journey, I acknowledge, accept, and embrace the subjectivity of my work, in accordance with Glesne and Peshkin's (1992) assertion:

> My subjectivity is *the* basis for the story that I am able to tell. It is a strength on which I build. It makes me who I am as a person *and* as a researcher, equipping me with the perspectives and insights that shape all that I do as researcher, from the selection of topic clear through to the emphases I make in my writing. Seen as virtuous, subjectivity is something to capitalize on rather than to exorcise. (italics in original, p. 104)

In the wake of this transition, moral autopilot gave way to an examination of my personal moral orientation. A pluralistic perspective regarding right and wrong, and good and bad emerged, which circumstantially and to varying degrees embraces virtue ethics, objective principles, care ethics, and consequentialism. These ethical theories are briefly reviewed, not only as they inform the lens of this book but also as they underpin the different approaches to moral education that are discussed in the chapters that follow.

Virtue ethics, also known as neo-Aristotelian virtue ethics, focuses on stable character traits or dispositions, which are categorized as virtues and vices. Virtues represent goodness, righteousness, and excellence and, therefore, are desirable. They enable one to act according to the highest potential of human goodness and to live a noble, admirable, and excellent life (Arthur, 2008; Carr, 1991, 2000; Markkula Center for Applied Ethics, 2006). Vices represent the antithesis and, thus, are undesirable. Both are considered universal, such that honesty, compassion, truthfulness, fairness, courage, moderation, and generosity are highly regarded in the literatures, customs, and norms of most world cultures (Fenstermacher, 2001).

As a normative theory of moral conduct, virtue ethics is agent-based (Carter, 2002). The moral status of a particular action or decision, therefore, is related to the character of the person performing that action or making that decision, regardless of the outcomes or perceived duties, obligations, and rights. Leming (2001) explains, "One's character is easily discerned by attending to one's habits. We say a person has good character when that person habitually acts in a manner determined to be virtuous" (p. 63). A person who habitually tells the truth and avoids lying, for example, is perceived to be an honest person. Additionally, a person who is reputed to be honest is generally assumed to be telling the truth and not lying. Individuals are thought to acquire virtues and, thus, achieve a virtuous character by direct inculcation (Lickona, 1991; Wynne & Ryan, 1997) and by associating with virtuous others (Fenstermacher, 2001).

The theory of objective moral principles is derived from Kantian deontology. According to Kant, the rightness or goodness of an action lies in the nature of the act itself. When a rational and autonomous person identifies a right, duty, or obligation, acting in accordance is morally correct regardless of the outcome or agent's character. For example, it is generally considered the right of stakeholders to know the truth and, therefore, the duty and obligation of those who know the truth, to tell the truth. There must also be agreement that a particular action can be universally applied, either in similar circumstances or as a categorical imperative (Carter, 2002; Hinman, 2006).

In the late 1970s, Tom Beauchamp and James Childress proposed four objective moral principles that define one's rights, duties, and obligations to preserve individual dignity and worth, and better the human condition. Sometimes referred to as *the big four*, respect for autonomy, beneficence, non-maleficence, and justice have shaped the foundation of biomedical ethics. Autonomy denotes the right of individuals to make their own choices; beneficence refers to acting in the best interests of others; non-maleficence is the principle of doing no harm; and justice relates to ensuring fairness, equity, and equality (Carter, 2002; Goodman & Lesnick, 2001). Acting according to these principles is considered morally correct. Similarly, Sissela Bok (in Goodman & Lesnick, 2001) identified objective principles, which are so fundamental to group survival that they have been established in even the smallest of communities. They are organized in the following three categories: (a) positive duties regarding mutual support, loyalty, and reciprocity for tending to those who are vulnerable; (b) negative duties regarding the avoidance of harm and wrongdoing; and (c) norms of fairness and justice, including equality and the rejection of bearing false witness. These coincide with beneficence, non-maleficence, and justice, respectively. As a final example, the Golden Rule is also considered to be a universally binding moral objective. Paraphrased positively, as treat others in ways you want to be treated, or negatively, as do not treat others in ways you do not want to be treated, it finds expression in Judaism, Christianity, Islam, Confucianism, and Hinduism, as well as other Eastern religions and cultures (Damon, 2005; Goodman & Lesnick, 2001).

Embracing theories of virtue ethics and objective principles represents a rejection of moral relativism, which embodies notions that moral codes are contextualized according to time, place, culture, religion, and circumstance; that there are no universally agreed upon virtues and vices, only acceptable or unacceptable behaviours in particular situations; and that there is no objective right and wrong, only different points of view (Campbell, 2003; Carter, 2002; Goodman & Lesnick, 2004). Although moral relativism has generally fallen out of favour in education discourse, I clarify my position, nonetheless, to circumvent the potential for confusion regarding three particular issues. First, the term *values* is often used synonymously with *virtues*, *morals*, and *ethics*. In this book, however, unless preceded by the adjective *moral* or *ethical*, as in *moral values* or *ethical values*, *values* are considered to be personal preferences or social constructs and conventions deemed desirable for particular communities, contexts, or situations. As such, values are subjective and relative and require inter-subjective understandings. Leming (2001) states, "When we speak of values we are referring to things or ideas that people hold dear. We may value many different things, from ice-cream to honesty. Some things we may value may have moral significance, some may not" (p. 63). Virtues and morals, therefore, are a subset of values, which are objectively and universally sanctioned.

Social-cognitive domain theory provides a framework for further understanding this distinction. Briefly, social-moral knowledge is constructed within three discrete domains—conventional, personal, and moral. The conventional domain is concerned with arbitrary yet agreed upon uniformities of social behaviour, which are determined by the social systems in which they are formed, including communities, organizations, religions, ethnic cultures, sports teams, societies, and schools. The personal domain is concerned with autonomy, individual prerogatives, and discretion. In this domain, the boundaries between self and others are delineated, and issues of self-control, personal choice, privacy, and identity are addressed. Finally, the moral domain is concerned with the well-being and welfare of others. It is categorical, objective, universal, and non-arbitrary, such that children and adolescents of diverse cultures, religions, and social classes, as young as 2.5 years of age, and with autism have demonstrated an ability to differentiate moral issues from conventional and personal issues (Nucci, 2001, 2008, 2009; Nucci & Turiel, 2009). Values are associated with the conventional and personal domains. There is nothing inherently right or wrong, or good or bad regarding actions defined within these domains. Virtues, and values qualified as *moral* or *ethical*, are associated with the moral domain and, therefore, can be objectively evaluated as morally right or wrong, good or bad.

This book is situated within the moral domain and, as such, concentrates on values that are moral and ethical in nature. Yet, conventional and personal values, which are operationalized in schools and classrooms, are not discounted. They have the potential to impact the moral nature of relationships and the learning environment. For example, a morally positive impact might occur when values of friendship and safety are promoted. A morally negative impact might occur with values of competition, individual achievement, and obedience. Further, conventional values may, in fact, represent morals. For example, obedience and compliance might be articulated in terms of respect, fairness, and responsibility. Lastly, values that are regularly challenged or contravened may signal moral confusion or unethical expectations. This might be the case when disobedience and noncompliance, for example, promote care, compassion, or justice.

The second potential point of confusion regarding moral relativism is related to the idea of morality as a social construct. Derived from Piaget's developmental theory and rooted in the progressive education movement, constructivism promotes the notion that children construct knowledge, including moral knowledge, by interacting with their physical and social worlds (Hildebrandt & Zan, 2008). Yet, each child is not constructing morality, per se, but building on centuries of moral wisdom to generate understandings of how virtues and morals are applied and enacted in social situations. Adults, as stewards of moral wisdom, help children to structure their particular experiences in a larger context of humanity (Damon, 2005; Hildebrandt & Zan, 2008).

Constructivism is at the heart of several contemporary approaches to moral education. Approaches such as cognitive development and care ethics, which promote an active role for teachers to reinforce objective and universal virtues and morals, do not generally foster moral relativism (Hildebrandt & Zan, 2008; Noddings, 2002). Approaches that recommend teachers remain neutral, such as values clarification and ethical reasoning, however, tend to foster moral relativism by equally valuing all viewpoints that students put forth, moral or otherwise. These latter approaches have, for the most part, been discredited as moral education (Balch, Saller, & Szolomicki, 1993).

Finally, nuances, complexities, and conflicts related to expressing, interpreting, and applying virtues and morals are not understood as moral relativism and do not justify a relativist argument. One teacher, for example, may display fairness by treating all students the same. Another teacher may display fairness by differentially accommodating students' individual needs. Among some cultural groups, respect is demonstrated by avoiding eye contact with persons of authority. Among others, respect is demonstrated by maintaining eye contact. Yet, fairness and respect are unlikely to be disputed as desirable and worthy moral values. Some situations place moral values in conflict with each other, creating ethical dilemmas for which there is no definitive right or good response. For example, honesty may conflict with compassion and sensitivity in situations that involve keeping one's confidence and avoiding hurtful truths; care may be at odds with justice and fairness when a struggling student receives a particular accommodation that is not offered to all students or, conversely, when such a student does not receive a special accommodation because it cannot be offered to all; and liberty and autonomy may conflict with loyalty and respect for authority. Yet, the goodness and rightness inherent in honesty, compassion, sensitivity, care, justice, fairness, liberty, autonomy, loyalty, and respect are not disputed as being objectively worthwhile and compatible with human flourishing, in the same way that harming, manipulating, neglecting, cheating, and intimidating others are not disputed as being objectively wrong and incompatible with human flourishing (Campbell, 2003; Carr, 2000, 2008; Nash, 2002; Noddings, 2002). As Hinman (in Goodman & Lesnick, 2001) contends, we may retain "sensitivity to the contextuality of our moral beliefs and the recognition that moral disagreement and conflict are permanent features of the moral landscape", but we also retain "the belief that some moral positions are better than others" (p. 100).

My pluralistic ethical perspective is also informed by care ethics, or the ethic of care, as it is also known. With roots in the modern feminist movement of the 1970s, this philosophical orientation emerged simultaneously from the fields of psychology (Gilligan, 1982) and philosophy (Noddings, 1984) in response to the male-oriented cognitive development theory. Advanced by Lawrence Kohlberg in the 1960s, cognitive development theory narrowly proposed that moral judgements, choices, and decisions are made rationally and in the pursuit of justice. Care ethicists, recognizing the female proclivity for caregiving in attachment relationships, assert that moral judgements, decisions, and choices may also entail impulse, attitude, and emotion, as well as the subjective.

As a normative theory of moral conduct, care ethics is situation-based and, thus, distinguished from a virtues conception of care. Noddings (in Goodman & Lesnick, 2001) explains:

> It is not properly labelled an ethic of virtue. Although it calls on people to be carers and to develop the virtues and capacities to care, it does not regard caring solely as an individual attribute. It recognizes that part played by the cared-for. It is an ethic of relation. (p. 73)

Hence, caring relationships between those who are cared for and their carers are necessarily reciprocal. The cared-for communicate their needs to carers and provide feedback regarding fulfillment of these needs. Carers are accountable for responding to the needs and feedback in a manner that is attentive, receptive, and nonjudgmental. These expectations involve the expression of a range of moral values, including honesty, respect, kindness, empathy, care, sensitivity, compassion, and trust. Thus, the preservation of reciprocally caring relationships is prioritized as morally good and right (Gilligan, 1982; Noddings, 2002, 2008).

Finally, I sometimes defer to a consequentialist perspective. Consequentialism represents several normative ethical theories that determine the moral status of an action or decision by the impartial evaluation of its consequences. A morally correct action or decision produces good and right consequences. One that is morally incorrect produces bad and wrong consequences. For example, lying may be considered right or wrong, good or bad. In a situation where lying protects one's feelings without harming others, it is morally right and good to lie. When lying harms one's reputation and does not provide benefit to others, it is morally wrong and bad. Consideration may also be given to the consequences of not acting, neither lying nor telling the truth, for example.

Further, consequences may be evaluated according to *benefit maximization*, a term used by Strike and Soltis (2009) in reference to achieving the greatest good for the greatest number of people. Utilitarianism, in particular, posits that the moral status of an action is based on its contribution to overall utility, variously understood as welfare, well-being, pleasure, and happiness (Carter, 2002; Driver, 2009; Hinman, 2006; Strike & Soltis, 2009). By way of example, utilitarianism provides a moral rationale for avoiding group punishment as a punitive and classroom-management strategy when only a few students have been disruptive. With most of the students punished without cause, the potential harm outweighs the potential good. As with all consequentialist theories, the character of the person who is acting, the principles or duties by which an action is performed, and relationships among stakeholders are not evaluated.

Although embracing four theories, this ethical perspective is not exhaustive. Of particular note, my lens excludes ethical orientations stemming from critical theory. In this theory, concepts of Marxism, social power, and social justice are applied to critically examine the role of authority, political structures, norms, and systems of society and schooling, as they undermine the rights, freedoms, and opportunities of non-dominant groups. This book describes the classroom practices of a teacher as they are enacted within existing systems and structures. But the systems and structures themselves are not critically examined. Still, the pluralistic framework enables me to cast a wide net regarding the moral dimensions of teaching, learning, and classroom life (Goodman & Lesnick, 2001). As Sanger and Fenstermacher (2000) observe:

> Any one normative theory fails to capture all that may be morally salient in a given context. Thus, the competitive, monotheoretic model of normative ethics is a poor one for the inquiries into [moral dimensions of teaching] because any one normative theory will only pick out, or focus on, say a benevolent character or a caring relationship or a universal principle of reason as morally salient, leaving other potentially salient features of the context hidden or less richly described. (p. 2)

THE STUDY

"What teacher is going to let you in the classroom for a *whole year*?" This honest challenge was posed by a fellow doctoral candidate and elementary-middle schoolteacher.

"That is precisely why the study is so important", I countered. "We need more access to classroom life, in real time and over a prolonged period of time, if we are going to fully appreciate the work that teachers do and share it around". In expressing this conviction, I fully committed, perhaps too naively, to the challenges and potential risks of a single-case, micro-ethnographic study.

Finding a participant who was willing and able to similarly commit to the project and who possessed the characteristics and qualities of interest was crucial. The following criteria were established to guide the selection process:

1. Full-time, classroom-based kindergarten to grade-12 schoolteacher;
2. Minimum three years of full-time teaching experience in a school;
3. Ethical knowledge and attributes of moral agency, including aware-ness of the moral dimensions of classroom life, one's ethical conduct and behaviours, and one's personal sense of morality as applied to teaching practices (Campbell, 2003);
4. Desire and effort to further the moral development of students;
5. Availability and willingness to collaborate throughout the school year;
6. Articulate communication skills, to facilitate collaboration.

The school context was not prioritized. Sean Patrick (pseudonym), a school administrator whom I had previously met, recommended an expe-rienced grade-four teacher at his own school. Along with homeroom responsibilities related to her students' overall well-being, Terry Kennedy (pseudonym) delivered curricula in language arts, math, health, science, and social studies, and supervised lunch three days per week. At the school level, she monitored outdoor recess twice on Tuesdays, coached the field hockey team, co-led the Peacemakers Program, and was the coordinator and liai-son for community outreach. She also taught a three-day back-to-school workshop on study skills the week prior to the start of school and volun-teered one night per week, programming and supervising children living in a nearby shelter who use the school's gymnasium for recreation.

In June of 2009, Terry and I met in her classroom for the first time. Our open and friendly conversation covered many relevant topics, including her teaching philosophy, vision, and priorities, all of which divulged a strong commitment to moral education and an understanding of the moral dimen-sions of teaching, learning, and classroom life. While interested in the proj-ect, Terry, nonetheless, expressed some honest concerns about participating. "I think the part that worries me most is the possible negative side effects. So let's discuss possible scenarios further. Also, I admit that it will be a little intimidating at first, to have you as part of our classroom on a regular basis. However, I am open to the experience. I know that I am not the perfect moral agent. There will be things I miss doing and missed opportunities, for sure".

Meeting for coffee at the beginning of August allowed me to address these concerns. "You were recommended for this study because you are successful at helping the students learn about right and wrong. My job will be to learn from you. You will also be helping me to interpret what I think I'm seeing. My intention is not to jump to conclusions without checking first with you".

"What if I yell at the kids one day?" Terry asked.

"I'm a mom. I have kids. Most days I can hold it together, but I sometimes yell and behave in ways that I'm not very proud of. It happens no matter how hard we try not to let it and no matter how much we know we shouldn't. I get that".

"How will you record it?" she pressed.

"You and I will decide together. Likely, I'll note it, and then in an interview I'll ask you what happened and why. We'll talk about it. Maybe we'll even see a pattern that will be helpful". I realized that I was pursuing Terry as a participant. The criteria were satisfied early on, but I was further impressed by her modesty and humility in light of a glowing recommendation from Sean. This is not an insignificant observation, as being a moral person is an essential aspect of moral agency. Hence, I persisted, "I'm not looking for perfection. There is no such thing. That wouldn't help anyone, anyway".

Finally convinced, Terry replied, "I hope I can give you what you need".

"I hope I can recognize and properly represent what you give me", I said with relief. Terry's curiosity and natural inclination toward learning overshadowed her self-consciousness.

From the first week of school, in September of 2009, until the last, in June of 2010, I spent a minimum of two full days per week immersed in Terry's classroom; I also accompanied the class on field trips and to special events and activities. Data collection methods included observation, artefact analysis, and interviews. My particular method of observation aligns with Fraenkel and Wallen's (2003) description of *observer-as-participant.* Primarily, I observed Terry and recorded field notes, often from the back of the classroom. Secondarily, I assisted Terry and the students in a variety of classroom activities and academic tasks, sometimes at the request of students and sometimes at Terry's request. Terry endorsed my participation by routinely announcing, "Gillian is someone you can also ask to help you". Toggling between participation and observation, or involvement and detachment, provides researchers with two useful perspectives. Participation allowed me to experience classroom life from the teacher's perspective, gaining an emic or insider's view (Fraenkel & Wallen, 2003). This was particularly significant, as I have never been a teacher of children and likely risked an outsider's tendency toward making uninformed assumptions. My sensitivity regarding how difficult it is for teachers to realize and maximize the potential morality of each situation, while also attending to the academic program, and many other facets of school and classroom life was greatly enhanced. At times, I felt overwhelmed by the bombardment of demands that students had for me, by the constant shifting of activities, and by a haunting feeling of unfinished business.

Some researchers, particularly those who conduct ethnographic anthropological studies, identify the risk of *going native*, or completely assuming an insider's role and becoming part of the community under study. Glesne and Peshkin (1992) warn, "The more you function as a member of the everyday world of the researched, the more you risk losing the eye of the uninvolved outsider" (p. 40). This risk has also been identified in the context of classroom-based micro-ethnographic studies (McCadden, 1998; Vazir, 2004). It was not difficult for me to avoid, however, as I felt no alignment with the role of teacher. Terry also maintained firm control of her class. I was not asked to make decisions, problem-solve, or plan; I was not expected to take initiative regarding the academic program, individual activities, or classroom life; and I was not left in charge. I was simply an extra pair of eyes, ears, and hands, directed by Terry and ever following her lead. Occasionally a student would ask how my research was progressing, suggesting I was successful in maintaining this separation and remaining marginal, with guarded intimacy (Glesne & Peshkin, 1992).

Also considered observational, I conducted an analysis of classroom artefacts, including posters, furnishings, decorations, worksheets, assignments, tests, student work, planning materials, syllabi, to-do lists, and reference books. Such artefacts are not usually created or utilized for moral reasons, per se, but may, nonetheless, have moral meaning and support a moral education agenda. This was the case, in fact, regarding several. They are noted in the chapters that follow, as they illustrate and detail the portrait.

Formal and informal interviews afforded access to Terry's thoughts, feelings, reflections, understandings, assumptions, vision, and motivations, as they define and guide her professional practice, generally, and her actions and reactions, more specifically. Eleven formal interviews, approximately one per month, were privately conducted, each lasting between 45 and 90 minutes. Semi-structured, they were guided by an overall theme and a series of open-ended questions. Yet, the dialogue was not overly regulated and often moved about in unanticipated and surprisingly informative directions. While affording an opportunity to deeply and collaboratively reflect on classroom life and teaching practice, such interviews were less informative when seeking clarification on particular events or incidents, due to the limitations of memory and hindsight. Once my position in the classroom was established, I also seized in-the-moment opportunities to ask emergent questions, such as "What do you think the students understood about that?"; "How do you feel about Heather's response?"; or simply, "What was all that about?" Eventually, Terry anticipated my curiosity, spontaneously offering editorial comments or initiating conversations when the students left the room. These informal interviews generated valuable insights, which were often revisited during a formal interview.

Last, Terry's practices are enacted in a broader school environment and to some degree are reliant on its particularities. Thus, historical and contextual information about the school are both helpful and necessary. Accordingly, I conducted a formal interview with Sean, who recounted the history of the school's Character Development Program and Terry's leadership role during its creation. School level artefacts, such as the school's website, handbooks, memos, newsletters, strategic plans, curriculum guides, and marketing brochures were also examined for implicit and explicit moral content. A more complete accounting of research activities, including protocols and schedules, is provided elsewhere (Rosenberg, 2013).

ROADMAP

Several themes have been introduced in this chapter to be further developed in the chapters that follow. The body of empirical, practical, philosophical, and theoretical work on moral education; the moral dimensions of teaching, learning, and classroom life; and moral agency is reviewed in Chapter Two, with citations primarily from the United States, Canada, Britain, Switzerland, Finland, and Sweden. This review establishes a theoretical framework by abstracting and repositioning understandings, insights, and knowledge according to four domains of teaching practice—teachers' vision; content to be taught and learned; teaching strategies; and methods of assessment. Chapters Three, Four, and Five address these domains by presenting analytic and anecdotal accounts of the research setting. Chapter Three situates Terry's professional practices in the contexts of her school environment and personal ideology. Terry's vision for classroom life and students' growth, development, and well-being is derived from her ideology, and is discussed in that section. Chapters Four and Five draw the portrait of moral agency, as teaching morally and teaching morality respectively, and detail content, strategies, and assessment methods related to the moral messages and lessons Terry imparts to students. The final chapter returns to the core concept of moral agency and suggests how this portrait might inform teaching practice, teacher education, and academic research.

At no point, do I intend to present a heuristic device or blueprint of moral agency, or to offer prescriptive recommendations for teachers, moral educators, or educators of educators. A single-case study does not serve that purpose, despite the consistency of these results with other empirical accounts. Rather, this portrait serves as an exemplar of what is real and possible in a particular context. My hope is that readers will find applications to their own situations, and be stimulated, inspired, and motivated to continue exploring these themes in practice and research, as well as discussing them with colleagues. While I believe that the broad themes presented will hold true because they are grounded in fundamental teaching pedagogy and teachers' folk wisdom, I also suspect that more than a single portrait can be drawn and that teachers might strive to find expressions of moral agency for themselves and to create their own unique portraits.

REFERENCES

Arthur, J. (2008). Traditional approaches to character education in Britain and America. In L. P. Nucci & D. Narváez (Eds.), *Handbook of moral and character education* (pp. 80–98). New York: Routledge.

Balch, M. F., Saller, K., & Szolomicki, S. (1993). Values education in American public schools: Have we come full circle? *Association for Supervision and Curriculum Development, 94*, 34. Retrieved March 25, 2009, from http://www.eric.ed.gov/ERICDocs/data/ericdocs2sql/content_storage_01/0000019b/80/15/94/95.pdf.

Berkowitz, M. W., Battistich, V. A., & Bier, M. C. (2008). What works in character education: What is known and what needs to be known. In L. P. Nucci & D. Narváez (Eds.), *Handbook of moral and character education* (pp. 414–430). New York: Routledge.

Boostrom, R. (1998). The student as moral agent. *Journal of Moral Education, 27*(2), 179–190. Retrieved January 16, 2011, from http://proquest.umi.com.

Campbell, E. (2003). *The ethical teacher.* Maidenhead: Open University Press McGraw-Hill.

Campbell, E. (2004). Ethical bases of moral agency in teaching. *Teachers and Teaching: Theory and Practice, 10*(4), 409–428.

Campbell, E., & Thiessen, D. (2000, April). *Moral and ethical exchanges in classrooms: Preliminary findings.* Paper presented at the Annual Meeting of the American Educational Research Association, New Orleans, LA.

Campbell, E., & Thiessen, D. (2001, April). *Perspectives on the ethical bases of moral agency in teaching.* Paper presented at the Annual Meeting of the American Educational Research Association, Seattle, WA.

Carr, D. (1991). *Educating the virtues: An essay on the philosophical psychology of moral development and education.* London: Routledge.

Carr, D. (2000). *Professionalism and ethics in teaching.* London: Routledge.

Carr, D. (2008). Character education and the cultivation of virtue. In L. P. Nucci & D. Narváez (Eds.), *Handbook of moral and character education* (pp. 99–116). New York: Routledge.

Carter, L. (2002). *A primer to ethical analysis: Major ethical theories.* Queensland: Office of Public Policy and Ethics of the Institute for Molecular Bioscience of the University of Queensland. Retrieved October 1, 2007, from http://www.uq.edu.au/oppe/PDFS/Ethics_primer.pdf.

Damon, W. (2005, Spring) Good? Bad? Or none of the above? *Education Next, 4*(2), 20–27. Retrieved September 13, 2010, from http://educationnext.org.

Davis, M. (2003). What's wrong with character education? *American Journal of Education, 110*(1), 32–57. doi: 10.1086/377672.

Dewey, J. (1959). *Moral principles in education.* New York: Greenwood Press.

Driver, J. (2009, Summer). The history of utilitarianism. In E. N. Zalta (Ed.), *The Stanford Encyclopedia of Philosophy.* Retrieved January 25, 2010, from http://plato.stanford.edu/archives/sum2009/entries/utilitarianism-history/.

Fallona, C. (2000). Manner in teaching: A study in observing and interpreting teachers' moral virtues. *Teaching and Teacher Education, 16*, 681–695.

Fenstermacher, G. D. (2001). On the concept of manner and its visibility in teaching practice. *Journal of curriculum studies, 33*(6), 639–653. doi: 10.1080/00220270110049886.

Fenstermacher, G. D., & Osguthorpe, R. (2000, April). *The manner of teachers and the character of students: What distinguishes character education from The Manner Project?* Paper presented at the Annual Meeting of the American Educational Research Association, New Orleans, LA. Retrieved July 3, 2008, from http://www-personal.umich.edu/~gfenster/.

Fenstermacher, G. D., Osguthorpe, R. D., & Sanger, M. N. (2009, Summer). Teaching morally and teaching morality. *Teacher Education Quarterly, 36*(3), 7–19. Retrieved February 25, 2010, from http://vnweb.hwwilsonweb.com.

Fraenkel, J. R., & Wallen, N. E. (2003). *How to design and evaluate research in education* (5th ed.). New York: McGraw-Hill.

Gilligan, C. (1982). *In a different voice: Psychological theory and women's development*. Cambridge, MA: Harvard University Press.

Glesne, C., & Peshkin, A. (1992). *Becoming qualitative researchers: An introduction*. Toronto: Longman.

Goodman, J. F., & Lesnick, H. (2001). *The moral stake in education: Contested premises and practices*. New York: Addison Wesley Longman.

Goodman. J. F., & Lesnick, H. (2004). *Education: A teacher-centered approach*. Boston: Pearson Education.

Hildebrandt, C., & Zan, B. (2008). Constructivist approaches to moral education in early childhood. In L. P. Nucci & D. Narváez (Eds.), *Handbook of moral and character education* (pp. 352–369). New York: Routledge.

Hinman, L. M. (2006). *Ethics updates*. University of San Diego. Retrieved June 19, 2009, from http://ethics.sandiego.edu/.

Howard, R. W., Berkowitz, M. W., & Schaeffer, E. F. (2004). Politics of character education. *Educational Policy, 18*(1), 188–215. doi: 10.1177/0895904803260031.

Joseph, P. B., & Efron, S. (2005, March). Seven worlds of moral education. *Phi Delta Kappan,* 525–533.

Kohn, A. (1997, February). How not to teach values: A critical look at character education. *Phi Delta Kappan,* 429–439. Retrieved February 15, 1020, from http://tigger.uic.edu/~lnucci/MoralEd/articles/kohn.html.

Leming, J. S. (2001). Historical and ideological perspectives on teaching moral and civic virtue. *International Journal of Social Education, 16*(1), 62–76. Retrieved March 17, 2011, from http://vnweb.hwwilsonweb.com.

Leming, J. S. (2008). Research and practice in moral and character education: Loosely coupled phenomena. In L. P. Nucci & D. Narváez (Eds.), *Handbook of moral and character education* (pp. 134–157). New York: Routledge.

Lickona, T. (1991). *Education for character: How our schools can teach respect and responsibility*. New York: Bantam Books.

Lickona, T., & Davidson, M. (2005). *Smart & good high schools: Integrating excellence and ethics for success in school, work, and beyond*. Cortland, NY, Washington, DC: Center for the 4th and 5th Rs/Character Education Partnership. Retrieved January 3, 2009, from http://www.cortland.edu/character/.

Markkula Center for Applied Ethics. (2006). *A framework for thinking ethically*. Retrieved October 10, 2007, from http://www.scu.edu/ethics/practicing/decision/framework.html.

McCadden, B. M. (1998). *It's hard to be good: Moral complexity, construction, and connection in a kindergarten classroom*. New York: Peter Lang Publishing.

Nash, R. J. (2002). *Real world ethics: Frameworks for educators and human service professionals*. New York: Teachers College Press.

Noddings, N. (1984). *Caring: A feminine approach to ethics and moral education*. Berkeley: University of California Press.

Noddings, N. (2002). *Educating moral people: A caring alternative to character education*. New York: Teachers College Press.

Noddings, N. (2008). Caring and moral education. In L. P. Nucci & D. Narváez (Eds.), *Handbook of moral and character education* (pp. 161–174). New York: Routledge.

Nucci, L. P. (2001). *Education in the moral domain*. New York: Cambridge University Press.

Nucci, L. P. (2008). Social cognitive domain theory and moral education. In L. P. Nucci & D. Narváez (Eds.), *Handbook of moral and character education* (pp. 291–309). New York: Routledge.

Nucci, L. P. (2009). *Nice is not enough: Facilitating moral development*. Upper Saddle River, NJ: Pearson Education.

Nucci, L., & Turiel, E. (2009). Capturing the complexity of moral development and education. *Mind, Brain, and Education, 3*(3), 151–159. doi: 10.1111/j.1751–28X.2009.01065.x.

Richardson, V., & Fenstermacher, G. D. (2000). *The Manner in Teaching Project*. Retrieved July 29, 2008, from www.personal.umich.edu/~gfenster.

Richardson, V., & Fenstermacher, G. D. (2001). Manner in teaching: The study in four parts. *The Journal of Curriculum Studies, 33*(6), 631–637.

Rosenberg, G. R. (2013). *Portrait of moral agency* (Doctoral dissertation). Retrieved March 29, 2015, from https://tspace.library.utoronto.ca/handle/1807/35945.

Ryan, K., & Bohlin, K. E. (1999). *Building character in schools: Practical ways to bring moral instruction to life*. San Francisco: Jossey-Bass.

Sanger, M. G., & Fenstermacher, G. D. (2000, April). *Aristotle is great—but is he enough? Expanding the theoretical grounds for inquiries into the moral dimensions of teaching*. Paper presented at the Annual Meeting of the American Educational Research Association, New Orleans, LA. Retrieved July 3, 2008, from http://www-personal.umich.edu/~gfenster/.

Setran, D. P. (2003, August). "From morality to character": Conservative progressivism and the search for civic virtue, 1910–1930. *Paedagogica Historica, 39*(4), 435–456. doi: 10.1080/00309230307474.

Sizer, T. R., & Sizer, N. F. (1999). *The students are watching: Schools and the moral contract*. Boston: Beacon Press.

Sockett, H. (2012). *Knowledge and virtue in teaching and learning: The primacy of dispositions*. New York: Routledge.

Stengel, B. S., & Tom, A. R. (2006). *Moral matters: Five ways to develop the moral life of schools*. New York: Teachers College Press.

Strike, K. A., & Soltis, J. F. (2009). *The ethics of teaching* (5th Ed.). New York: Teachers College Press.

Vazir, N. A. (2004). *Learning about right and wrong: Perspectives of primary students of class 1 in a private school in Pakistan*. (Doctoral dissertation, University of Toronto, Toronto).

Watson, M. (2008). Developmental discipline and moral education. In L. P. Nucci & D. Narváez (Eds.), *Handbook of moral and character education* (pp. 175–203). New York: Routledge.

Weissbourd, R. (2003, March). Moral teachers, moral students. *Educational Leadership, 60*(6), 6–11.

Wynne, E. A., & Ryan, K. (1997). *Reclaiming our schools: Teaching character, academics, and discipline*. Upper Saddle River, NJ: Prentice-Hall.

2 The Shoulders I Stand On
Theoretical Framework

It is widely accepted that schooling in Western society has always comprised a moral component (Howard, Berkowitz, & Schaeffer, 2004; Sockett, 2012; Straughan, 1988). What constitutes moral education and how overtly it is expressed in public schools and classrooms, however, has evolved, reflecting and responding to changing religious, political, sociological, cultural, psychological, philosophical, and technological contexts. In particular, at the end of the Second World War the influence of logical positivism and the growing need for high-level technical skills among workers compelled a science-based agenda. This required increased class time for academic and cognitive learning. Explicit forms of moral education were withdrawn from classrooms, and by the early 1950s moral learning was relegated to home and religious life (Arthur, 2008; Balch, Saller, & Szolomicki, 1993; Beachum & McCray, 2005; Field, 1996; Goodman & Lesnick, 2001; Hunt & Mullins, 2005; McClellan, 1999).

Toward the latter part of the 1960s, North American society was perceived by some to be in moral disarray. Ryan (1988) concluded, "What is clear is that there has been an explosion of anti-social behaviour among the young since World War II" (p. 18). Many attributed this to the decline of direct moral instruction in public schools. Hunt and Mullins (2005) are among them, noting, "The impact of such devaluation of moral education was felt with devastating results in American public schools" (p. 182). Postwar sociopolitical turbulence was also accused. The space and arms races between the Soviet Union and the United States, the Red Scare, the Korean Conflict, the Cuban Missile Crisis, the Vietnam War and antiwar demonstrations, the women's movement, the civil rights movement, the sexual revolution and widespread use of the birth control pill, the rising rate of divorce and altered family structures, immigration and demographic changes, and the increased use of illegal recreational drugs all contributed to diversified notions of race, equality, and public morality, such that home, community, school, and religion no longer seemed to project a unified or consistent moral code. Balch, Saller, and Szolomicki (1993) note that people were "bewildered and confused about how they should feel and behave" (p. 12). Further, the rising concern for individual autonomy, freedom and rights, and a general distrust of authority challenged traditional relationships between students and teachers, and ideals of classroom life. Many teachers reevaluated their professional practices. Fearing accusations of indoctrination and manipulation, they began to operate more as technicians dispensing information than professionals nurturing the holistic growth and development of children (Balch, Saller, & Szolomícki, 1993; Beachum & McCray, 2005; Cohn & Kottkamp, 1993; Field, 1996; Howard, Berkowitz, & Schaeffer, 2004; McClellan, 1999). Finally, the unrestricted and far-reaching emergence of media-driven popular culture was vastly influencing children's developing value systems (Lickona, 1991; Sizer & Sizer, 1999) and exposing them to values that were considered "largely antithetical to intellectual and moral excellence" (Lickona & Davidson, 2005, p. 11).

Teachers were "called upon to offer a counterweight to the malformative elements permeating children's lives" (Narváez & Lapsley, 2008, p. 158). Between the mid-1960s and late 1990s, several prescribed school-based approaches to moral education were attempted, to varying degrees of success. The most notable are values clarification, ethical reasoning, character education, and applications of cognitive development and care ethics theories (Balch, Saller, & Szolomicki, 1993; Howard, Berkowitz, & Schaeffer, 2004; Joseph & Efron, 2005). Yet, by the end of the century widespread concern persisted, and criticisms were levelled at public schools for failing to effectively impart morality to students (Lickona, 2004; Ryan & Bohlin, 1999; Schuitema, ten Dam, & Veugelers, 2008; Wynne & Ryan, 1997). Stengel and Tom (2006) recall, "Public concern grew about teen pregnancy, school violence, and other social ills, and a steady stream of articles appeared in newspapers, magazines, and professional journals about the need for more attention to the moral aspect of schooling" (p. 17).

Educators, academics, and parents continued to speculate that children learn about right and wrong in school and from their teachers. However, there was limited empirical evidence to illuminate how this might be accomplished. From the early 20th century, education research, as advanced by behavioural psychologist Edward Thorndike, was largely influenced by a paradigm derived from the physical sciences (Fenstermacher, 2001; Leming, 2008). Rooted in objectivism, positivism, and determinism, this paradigm held the following assumptions: (a) the social world, like the natural world, is governed by universal laws that regulate human behaviour and interaction; (b) reality, truth, and knowledge are independent of humans and can be captured, understood, and verified; and (c) researchers are objective and autonomous observers of social reality. Accordingly, research was deductive and quantitative in nature, typically involving surveys that measure cause-and-effect relationships, test hypotheses, control or remove confounding variables, and nullify biases and assumptions. Results were objectively represented and generalizable (Cohen & Manion, 1994; Fraenkel & Wallen, 2003). Cohen and Manion (1994) observe, however, "No matter how exact measurement may be, it can never give us an experience of life, for life cannot be weighed and measured on a physical scale" (p. 23). Thus, many believe this methodology limited the scope of education research and was an obstacle to understanding and informing practices that characterize teaching and facilitate learning (Leming, 2008, 2010).

Education research was eventually influenced by the work of philosopher and psychologist John Dewey. A contemporary of Thorndike, Dewey (1929) believed that education was as much an art as a science. This paved the way for qualitative research methodologies. Denzin and Lincoln (2000) note, "The province of qualitative research, accordingly, is the world of lived experience, for this is where individual belief and action intersect with culture" (p. 8). Inductive, naturalistic, and constructivist, this research paradigm considers the social world to be a manifestation of human consciousness, behaviour, and interaction; and considers reality, truth, and knowledge to be contextually created and experienced. As such, researchers are not assumed to be entirely objective or autonomous; results are not often generalizable; and variables, biases, and assumptions cannot always be controlled (Cohen & Manion, 1994; Denzin & Lincoln, 2000; Fraenkel & Wallen, 2003; Howe & Eisenhart, 1990; Krauss, 2005). Yet, observations, interviews, and examination of artefacts, the main sources of qualitative data, made possible new insights and perspectives on the work of teachers, and the school and classroom experiences of students.

Empirical investigation of moral education, more particularly, was redirected in the early 1990s by the ground-breaking book entitled *The Moral Dimensions of Teaching* (Goodlad, Soder, & Sirotnik, 1990). This book's range of topics, including teacher professionalization and professionalism, the moral purpose of compulsory schooling, accountability, the moral dimensions of the classroom, moral relationships between teachers and students, and ethical codes provided both a theoretical foundation for investigating moral education in schools and classrooms (Campbell, 2008b) and the justification for using qualitative methodologies to do so. Six notable qualitative studies were conducted in the wake of this publication. The Moral Life of Schools Project (Jackson, Boostrom, & Hansen, 1993) represents breakthrough work on the moral dimensions of the classroom. For two-and-a-half years, investigators observed 18 classrooms in six Midwest United States schools. Their data identified eight categories of moral influence by which teachers transmit moral messages and lessons to students. The Manner in Teaching Project (Fenstermacher, 2001; Richardson & Fenstermacher, 2000, 2001), also from the United States, assumed an Aristotelian virtue ethics perspective. Conducted over three years, this study involved 11 teachers, two principals, and several students in two distinctly different Michigan schools. With interviews, videotapes, group meetings, and observations, investigators detected six methods by which teachers nurture students' moral and intellectual virtues. These eight categories and six methods are outlined in the strategies section below. Independent articles from both studies are also cited in this chapter and elsewhere in the book, including Hansen (2002), from The Moral Life of Schools Project, as well as Fallona (2000), Fenstermacher and Osguthorpe (2000), Richardson and Fallona (2001), Sanger (2001), and Sanger and Fenstermacher (2000), from The Manner in Teaching Project.

The Moral and Ethical Bases of Teachers' Interactions with Students study (Campbell, 2003, 2004; Campbell & Thiessen, 2000, 2001) investigated the moral agent role of 12 teachers in five diverse Canadian urban schools. Observations, artefact analysis, and interviews focused on classroom rules, routines, and norms of behaviour; teaching and learning materials; pedagogical approaches; and methods of evaluation and assessment. The results revealed a range of moral values that are knowingly expressed and prioritized in the professional conduct, behaviours, and practices of these teachers, most notably fairness, respect, kindness, and honesty. These four moral values help to structure the discussion of Terry's expressed morality in Chapter Four. Simon's (2001) Moral Questions in the Classroom: How to Get Kids to Think Deeply about Real Life and Their Schoolwork documents a three-month study, conducted in three high schools in the United States. Classroom observations and teacher and student interviews determined that American high school curricula provide ample opportunities for teachers and students to engage in moral, spiritual, and existential exploration of issues related to humanity, society, and their own lives. Yet, the teachers tended to discourage doing so in the public forum of the classroom, instead relegating such exploration to individual homework assignments. Simon's thoughts on utilizing academic curricula for moral education are cited in the strategies section below.

The study entitled It's Hard to be Good: Moral Complexity, Construction, and Connection in a Kindergarten Classroom (McCadden, 1998) is methodologically similar to my own work. The daily life of a North American public school class is explored over an entire school year, with methods of participant and non-participant observation, and interviews with the teacher, several students, and a teacher's aide. Two distinct cultures through which students rotate during their school day were identified: the organizational culture inside the classroom, which is created and managed by the teacher, and the relational culture on the playground, which is created and managed by students. Finally, The Moral Dimensions of Teaching: Language, Power, and Culture in Classroom Interaction (Buzzelli & Johnston, 2002) applied the tripartite framework of language, power, and culture to the discourse analysis of classroom transcripts and to the content analysis of course syllabi. Particular forms of discourse, the use of specific language, teachers' choices with respect to the register of their voice, notions of teachers' authority, both as *an* authority and *in* authority, and the ethnocentricity of a dominant culture were identified as carrying moral meaning in social interactions between teachers and students.

By the beginning of the 21st century, a substantive body of empirical, practical, philosophical, and theoretical scholarship on moral education; the moral dimensions of teaching, learning, and classroom life; and moral agency had accumulated. These works represent the shoulders on which I stand. Anticipating the exploration of Terry's moral agency, I pass the knowledge and insights they have generated through an applied lens and reframe the literature according to four domains of teaching practice—teachers' vision; content to be taught and learned; teaching strategies; and methods of assessment. The remainder of this chapter outlines what is understood as moral within each domain.

TEACHERS' VISION

Kosnik and Beck (2009) characterize teachers' vision as a "vast network of ideas, principles, and images touching on both theory and practice" (p. 174). As such, vision seems to fall somewhere between philosophy and beliefs, and knowledge and plans, embodying the professional and personal desires, dreams, aspirations, hopes, and cares that individual teachers hold for themselves, their classrooms, and their students. Both cognitive and emotional, vision shapes teachers' attitudes and objectives; influences their interactions with students; informs the classroom's culture, routines, activities, and physical space; and provides a yardstick by which teaching and learning might be measured (Hammerness, 1999, 2006). As Kosnik and Beck (2009) imply, vision functions as an adhesive of sorts:

> Part of having a vision is understanding *how the various aspects of teaching fit together*; not just the activities within a particular curricular area (such as literacy) but how the program as a whole, the classroom management strategies, the assessment methods, and the type of community form an *approach* to teaching (italics in original, p. 170)

It is not clear how such visions are formed. Teachers' cumulative life experiences, both personal and professional, seem to be of significant influence (Campbell, 2003). This includes family life, relationships, religious beliefs and practices, hobbies and interests, prior employment, and the current teaching context. Although steadfast ideals may develop, teachers likely learn, through experiences and experimentation, what is achievable or unfeasible, supported or obstructed in particular school environments, and are able to modify their visions accordingly (Hammerness, 2002, 2006).

In many cases, teachers' vision is moral in nature. Campbell (2003) reports that some teachers can "articulate with depth and intention what they hope to achieve morally and ethically in their classrooms and how they hope to facilitate it" (pp. 38–39). Buzzelli and Johnston (2002) report that some teachers strive to nurture the people their students might become as productive members of society. Ryan and Bohlin (1999) suggest that great teachers sustain a vision of their students' character potential. Coloroso (2005) claims that teachers envision not only what they would like their students to know about morality but also what they would like students to do with that knowledge. Finally, The Manner in Teaching Project (Fenstermacher, 2001; Richardson & Fenstermacher, 2000, 2001) revealed a more immediate moral objective concerning the classroom climate:

> All teachers in our study have a vivid conception of the kind of place they want their classrooms to be. Mutual respect, sharing, tolerance, orderliness and productive work are the notions most often mentioned by the teachers when describing their aspirations for their classrooms. (Fenstermacher, 2001, p. 642)

These examples illustrate that many teachers maintain a moral education agenda, at least conceptually.

To explore Terry's vision, I pose the question: What is the teacher's vision, as related to the moral education of students? Table 2.1 anticipates possible answers, according to the following: (a) the classroom environment, (b) student achievement, and (c) teacher identity. Each is associated with moral education, as follows. A classroom environment characterized by moral values of respect, care, and inclusiveness, among others, provides a context in which moral behaviours and attitudes are acquired and practiced. Student achievement denotes learning, growth, and development in a number of domains, including social-moral, that rely on expressing virtues and demonstrating moral reasoning. Teacher identity, or how the teacher envisions his or her professional role, responsibilities, and relationships with students, is related to the teacher as a moral person and a moral educator. These associations are taken up more thoroughly in the sections below as they become relevant to content, strategies, and assessment. Here, they are addressed as possible subjects of teachers' vision.

The classroom environment might be broadly envisioned as a democracy, according to metaphors of community, family, team, or tribe, or through the lens of a particular ethical orientation, such as utilitarianism or justice. More specifically, and not unrelated, teachers may refer to the classroom's physical space, organizational structures, or culture. Regarding physical space, Hammerness (2002) recalls one teacher who clearly imagined the bookshelves, windows, colour, and condition of walls, posters, artwork, and desk arrangement. Further, a teacher whose ideal class is described with the team metaphor might envision dedicated desk and storage spaces for each student, similar to a changing room with dedicated lockers for each member. A family metaphor may evoke more communal spaces. Classroom structures that might be featured in a teacher's vision include procedures for accountability, assessment, and discipline; processes for collaboration, sharing, decision-making, and problem-solving; routines for set-up, tidy-up, and relocation; and pedagogical approaches, such as lecturing, constructivist activities, group work, and service-learning (Campbell 2003; Howard, 2005;

Kosnik, Beck, Cleovoulou, & Fletcher, 2009; Hammerness, 1999, 2002). Describing an ideal classroom as a democracy might imply structures of collaborative decision-making and problem-solving. A utilitarian perspective may imply a focus on the efficiency of carrying out routines.

Finally, classroom culture refers to values that are espoused and/or expressed, norms of behaviour, shared beliefs and assumptions, and relationships among students. Fenstermacher (2001) identifies teachers' aspirations to instil conventional values of productive work, orderliness, and sharing. Hammerness (2002) recounts one teacher's desire for a culture of excitement, earnestness, and life. A justice orientation suggests moral values

Table 2.1 Vision

Classroom environment:	• democracy • metaphor (e.g., community, family, team, tribe) • ethical orientation (e.g., utilitarianism, justice)
physical space	• shelves, windows, posters, artwork, wall colour, desk arrangement, storage
structures	• procedures (e.g., accountability, assessment, discipline) • processes (e.g., collaboration, sharing, decision-making, problem-solving) • routines (e.g., set-up, tidy-up, relocation) • pedagogical approaches (e.g. constructivist activities, group work, service-learning)
culture	• values (e.g., productive work, orderliness, sharing excitement, earnestness, life) • moral values (e.g., fairness, equality, impartiality, empathy, compassion, kindness, sensitivity) • student relationships (e.g., friends, colleagues, partners, collaborators, teammates, members, co-investigators)
Student achievement:	• learning, growth, development, as underpinned by moral values and moral reasoning • academic (e.g., accountability, honesty, responsibility) • social (e.g., kindness, fairness, respect, peer mediation) • behavioural (e.g., autonomy, self-regulation, courage) • intellectual (e.g., impartiality, integrity, problem-solving, decision-making) • code of conduct, set of rules • short-term or long-term objectives
Teacher identity:	• role (e.g., leader, lead learner, inspirer, guide, facilitator, collaborator, resource, resource provider, purveyor of knowledge, carer) • relationship with students (e.g., confidant, friend, mentor, coach, surrogate parent, big brother or big sister) • responsibilities (e.g., ideals of professionalism, being good, trying one's best, the Golden Rule)

of fairness, equality, and impartiality, and care ethics suggests empathy, compassion, kindness, and sensitivity (Noddings, 2002). Further, relationships among students may be envisioned as friends, colleagues, partners, collaborators, teammates, members, and co-investigators (Hammerness, 2002). Each descriptor entails expectations of how students might interact— for example, friends on a more intimate level, with care, kindness, and respect, and colleagues, guided by objective principles of autonomy and fairness. A certain consistency is assumed regarding how the teacher envisions the physical appearance of the classroom, the structures that determine how tasks and responsibilities are undertaken, and the culture that defines a way of life, regardless of whether a metaphor or ethical orientation is articulated to contextualize these environmental elements.

Vision for student achievement encompasses what the teacher hopes for in terms of learning, growth, and development. This might involve academic, social, behavioural, and intellectual domains. Achievement in each is often correlated with the expression of moral values and moral reasoning. For example, the teacher may envision academic achievement in relation to accountability, honesty, and responsibility; social achievement as kindness, fairness, respect, and morally defensible peer mediation; behavioural achievement as autonomy, self-regulation, and courage; and intellectual achievement as impartiality, integrity, and morally correct problem-solving and decision-making (Coloroso, 2005; Hammmerness, 1999, 2002, 2006; Lapsley, 2008). Such a vision might be made explicit in a code of conduct or set of rules to which students aspire, or may remain implicitly understood as a desired outcome of schooling. Additionally, vision can represent short-term objectives for realization in the present context, or long-term objectives for whom students might become as good citizens and people of character (Buzzelli & Johnston, 2002; Campbell, 2003, 2008a, 2008b; Ryan & Bohlin, 1999). Regarding the latter, Campbell (2003) quotes one teacher, as follows:

> I'm planting the seeds, and the seeds will at some point in time in their lives, they'll blossom. Maybe not right now; maybe one student out of the 28 may get it now. Who knows, but I'm optimistic, and if I can reinforce in them the right behaviour, at some point in their lives, they'll get it. They'll understand. (p. 56)

Lastly, in this context teacher identity entails the teacher's perception of his or her professional role, relationships with students, and responsibilities. The teacher's role might be characterized as leader, lead learner, inspirer, guide, facilitator, collaborator, resource, resource provider, or purveyor of knowledge. Relationships, although embedded in these descriptors, may be more specifically envisioned as confidant, friend, mentor, coach, surrogate parent, or big brother or sister (Hammerness, 2002, 2006; Sanger, 2001). A teacher who articulates his or her role as guide or facilitator, for example, likely also envisions a mentor or coach relationship with students.

In addition, metaphors illustrating ideal classroom environments and teachers' ethical orientations reveal aspects of teacher identity. A family metaphor evokes the role of leader or purveyor of knowledge and the relationship of surrogate parent or big brother or big sister. A care ethics orientation, by definition, places teachers in the role of carer (Noddings, 2002, 2008, 2010). A virtue ethics orientation might indicate the role of guide or facilitator. Finally, teachers might express a vision of how they hope to carry out responsibilities, philosophically, with ideals of professionalism, being good, and trying one's best, or practically, as guided by the Golden Rule (Campbell, 2003; Sanger, 2001).

Education literature merely begins to identify the many possible permutations and combinations of classroom environments, student achievement, and teacher identity that might shape the moral and ethical objectives of teachers' vision. As ideals and aspirations, however, vision cannot be directly observed in classroom practices because it may not be fully achieved or fully achievable. The descriptions of Terry's vision in Chapter Three, therefore, rely heavily on interview conversations. Observations are used by way of illustration but not as a means of judging how well Terry's vision is realized or to indicate the status quo.

MORAL CONTENT

The term *content* broadly signifies what teachers convey to students as moral education. This includes both knowledge and abilities. Aligning with Aristotle's practical moral wisdom, knowledge content refers to understanding human life in terms of good and bad, and right and wrong, and understanding how this is applied in social behaviour. Abilities content refers to skills associated with the cognitive and emotional processing of morally salient issues. These two types of content are typically associated with different approaches to moral education: knowledge with character education, and abilities with cognitive development theory. It is not my intention to fully review these approaches or to confine this discussion in accordance with them, but rather to extract what each aims to convey to students as part of a wider consideration of moral content.

Rooted in normative virtue ethics, character education endeavours to cultivate and habituate moral values as character traits or dispositions. This traditionally entails delivering formal, didactic lessons and upholding conformity to standards of moral behaviour and good habits (Lickona, 1991, 2004; Ryan & Bohlin, 1999; Wynne & Ryan, 1997). Accordingly, character programs revolve around a set of preselected moral values, which tend to vary only slightly from program to program. For example, the Character Education Partnership (Character.org, n.d.), an advocacy and leadership coalition in the United States, offers programs and materials that focus on promoting justice, compassion, honesty, respect, responsibility, and diligence. The Character Counts! program (Josephson Institute, n.d.), hosted by the Josephson Institute of Ethics in California, endorses six pillars of character: trustworthiness, respect, responsibility, fairness, caring, and citizenship. Created by Live Wire Media, also in California, goodcharacter. com (n.d.) provides lesson plans, activities, and other resources emphasizing trustworthiness, respect, responsibility, fairness, caring, and citizenship for elementary level students, with the addition of justice, honesty, courage, diligence, and integrity for middle and high school levels. Lastly, the Giraffe Heroes Project (n.d.), based in Washington State, primarily promotes courage, as it honours the "risk-takers: people who are largely unknown, people who have the courage to stick their necks out for the common good, in the US and around the world". This is merely a sampling of literally thousands of commercially available programs that are identified as character education.

Additionally, character education programs may be locally developed to suit particular community needs. In many cases, this was encouraged or mandated by education policy. Ontario's Education Act is of particular relevance to this portrait. Toward the end of the 20th century, the Ontario Ministry of Education declared that all publicly funded school boards would be responsible for teaching values and that students should learn, at school, to demonstrate self-discipline, cooperation with others, care for others, social responsibility, citizenship, responsibility for their own behaviour, and thoughtful conflict-resolution strategies. In October of 2006, this culminated in the launch of a character development initiative, called *Finding Common Ground: Character Development in Ontario Schools, K-12* (Glaze, Zegarac, & Giroux, 2008). All publicly funded kindergarten to grade-12 schools were subsequently mandated to create and implement a character development program, with universal attributes that would permeate school life, bind stakeholders in a shared cause, and form the basis of responsible citizenship in a just and democratic society.

One of the earliest adopters of this initiative, the York Region District School Board (n.d.), developed the Character Matters program around 10 attributes: respect, responsibility, honesty, empathy, fairness, initiative, courage, integrity, perseverance, and optimism. These attributes are "interwoven through every aspect of school life, from how students and staff members greet one another to how literature and social studies are discussed, to expectations of conduct in sports" (York Region District School Board, n.d.). Acknowledging its 30% to 35% Aboriginal population, the Keewatin-Patricia District School Board (n.d.) selected seven Grandfather Teachings of Aboriginal culture, upon which their character program was built—bravery, responsibility, respect, kindness, honesty, humility, and wisdom. These attributes are reinforced in classroom and school-wide activities, and in a *pay-it-forward* initiative involving acts of kindness throughout the community. As a final example, the Rainbow District School Board (n.d.) nurtures 13 personal qualities by way of instilling essential skills for learning and life and helping students to become positive and productive members of a global society. These personal qualities include respect, responsibility, honesty, integrity, fairness, inclusiveness, teamwork, trust, initiative, perseverance, determination, optimism, and compassion.

Whether commercially or locally developed, such programs are often informed by the work of scholars who promote varying sets of moral values. For example, Lickona (1991) advocates for honesty, fairness, tolerance, prudence, self-discipline, helpfulness, compassion, cooperation, and courage. Stengel and Tom (2006) recommend six pillars of character—trustworthiness, respect, responsibility, justice, caring, and civic virtue. And Borba (2001) endorses seven virtues—empathy, conscience, self-control, respect, kindness, tolerance, and fairness. While Wynne and Ryan (1997) generally support this approach to moral education, they caution that a list of more than six to eight virtues will entail overlaps and likely not offer additional benefit.

Perhaps in acknowledgement of this observation, frameworks and hierarchies have been developed. Lickona (2004) extends his previous work to identify 10 virtues within two categories derived from Aristotle's concept of the *life of right conduct*. Right conduct toward oneself includes wisdom, fortitude, self-control, positive attitude, hard work, integrity, and humility. Right conduct toward others includes justice, love, and gratitude. Lapsley (2008) suggests Blasi's framework for nurturing self-identity through lower-order and higher-order virtues. Lower-order virtues of empathy, compassion, fairness, honesty, generosity, kindness, and diligence are associated with situation-specific responses. Higher-order virtues are more broadly applicable and organized according to two clusters. Willpower involves virtues of self-regulation and self-control in problem-solving. Integrity involves internal self-consistency, with respect to keeping one's word and being transparent, accountable, responsible, and sincere. Further, some hierarchies propose one or two *master* values through which a more complex moral orientation might be explored. Lickona (1991) and Lickona and Davidson (2005) promote respect and responsibility, noting, "Respect and responsibility are the 'fourth and fifth R's' that schools not only may but also must teach if they are to develop ethically literate persons" (Lickona, 1991, p. 43). Goodman and Lesnick (2001) promote integrity. Cooper (2010) promotes empathy. A former research participant of mine, who teaches middle school, promotes compassion, while another, who teachers high school math, promotes respect (Rosenberg, 2008).

Abilities content that might be included as moral education generally refers to the development of cognitive and emotional processes for understanding and judging issues of moral salience, and for making right and good choices and decisions. In the 1960s, American psychologist Lawrence Kohlberg advanced cognitive development theory, proposing six hierarchical stages of cognition by which such abilities might be nurtured and improved. Accordingly, moral education entails progressing students through these stages by harnessing natural and intrinsic developmental processes, and providing opportunities for personal discovery. In the mid-1970s, the emergence of social-cognitive domain theory, as developed by Nucci and Turiel (Nucci, 2008, 2009; Nucci & Turiel, 2009), responded to challenges levelled against Kohlberg's single developmental pathway by delineating three integrated pathways or domains—moral, conventional, and personal. Each was briefly outlined in Chapter One. Encouraging increasingly sophisticated and complex moral reasoning skills among students, however, remains the educational objective.

Moral psychologist James Rest identified more specifically what these reasoning skills or abilities might be. In The Four-Component Model of Morality, Rest (1986) proposed four psychological processes for completing a moral or ethical action and sustaining a morally good life: moral sensitivity, moral judgment, moral motivation, and moral character. Moral sensitivity is the ability to recognize and interpret moral salience in a given situation and to identify options or possible courses of action. Moral judgment or reasoning is the ability to determine which available options are morally necessary and acceptable. Moral motivation is the desire to implement the selected option, despite other inclinations or competing priorities. Moral character refers to the perseverance, persistence, and courage that might be required to implement the selected option when confronted with barriers, challenges, or obstacles. With this forth component, Rest recognizes the significance of one's character in a moral action. Similarly, character educators Wynne and Ryan (1997) recognize the significance of cognitive processes in cultivating and habituating virtue. Hence, they echo Rest with the following recommendations:

- Being able to think through questions concerning what is right and wrong;
- Sorting out facts related to situations of a moral nature;
- Connecting such situations to principles;
- Thinking through solutions for moral problems;
- Selecting the most ethical solution.

Cognitive development theory traditionally focused on reasoning abilities. Yet, with the work of Carol Gilligan (1982), among others, an affective component was added. Cognitive-affective interaction is implicitly implied in Rest's model, as one must be sensitive to the welfare of others and sustain a desire to act. Joshua Greene develops this further, however, in proposing a dual-process theory whereby moral judgments are shaped by cognitive processes, such as reasoning and self-control, as well as automatic processes, such as emotional "gut reactions". Both play a critical role in moral judgment. As Greene asserts, "Characteristically deontological moral judgments (judgments associated with concerns for 'rights' and 'duties') are driven by automatic emotional responses, while characteristically utilitarian or consequentialist moral judgments (judgments aimed at promoting the 'greater good') are driven by more controlled cognitive processes" (parentheses and quotations in original, Green, n.d.). Again, this is but a small sampling from the very large body of work, particularly in the field of psychology, that might inform moral education.

To explore the content of Terry's moral education for students, I pose the question: What is the content of the teacher's moral messages and lessons? Table 2.2 summarizes the discussion above and anticipates possible answers

Table 2.2   Content

Knowledge content:	• set of moral values	
	• framework or hierarchy of moral values	
	• master moral value(s)	
moral values	• kindness	• justice
	• love, care, compassion	• fairness
	• sensitivity, empathy	• equity
	• respect	• equality, impartiality
	• forgiveness, mercy	• responsibility
	• courtesy, thoughtfulness	• accountability, dependability
	• understanding	• diligence, persis-tence, perseverance, determination
	• sympathy	
	• sincerity, genuineness, authenticity	• courage, fortitude, bravery
	• tolerance, acceptance	
	• cooperation, helpfulness	• friendliness
	• inclusiveness	• prudence, discretion
	• humility, modesty	• moderation
	• integrity	• temperance
	• honesty, truthfulness	• hope, optimism
	• trustworthiness	• self-discipline, self-control, willpower, self-regulation
	• trust	
	• faithfulness, loyalty	
	• gratitude, appreciation, thankfulness	• beneficence, avoiding maleficence and harm
	• patience	
	• generosity, charity	
Abilities content:	• stage progression	
	• domain specificity (moral, conventional, personal)	
	• cognitive-affective interaction	
skills, processes	• reasoning, judging	• moral sensitivity
	• reflecting, thinking, con-sidering, contemplating	• moral motivation
		• identifying options
	• understanding, knowing	• choosing solutions and actions, decision-making
	• rationalizing, justifying	
	• evaluating	• gut reactions
	• resolving, problem solving	• attachment and disengagement

according to specific knowledge and abilities, and to how knowledge and abilities might be conceived and organized for the classroom. Knowledge content includes a list of more than 60 moral values, and indicates where overlaps may occur. The aim of this list is to be comprehensive, rather than concise. Three ways in which moral values might be organized for delivery are noted—as a set, framework or hierarchy, or master value. Abilities content also includes a comprehensive, rather than concise, list of processes and skills. They are primarily cognitive in nature but also include emotional abilities related to attachment, disengagement, and gut reactions. For moral education, abilities might be conceived according to stage progression, domain specificity, or cognitive-affective interactions.

STRATEGIES FOR IMPARTING MORALITY

Although closely connected, a distinction is made between moral content and moral instruction, according to Fenstermacher and Osguthorpe's (2000) assertion that "a teacher may be cultivating moral ends, and doing so intentionally, without specifically addressing moral content" (p. 4). Strategies, therefore, broadly encompass any means, methods, measures, modes, ways, approaches, practices, procedures, processes, techniques, and pedagogies by which teachers convey morality to students, purposefully or coincidentally, directly or indirectly, and formally or informally. Eight categories of moral influence were identified by The Moral Life of Schools Project (Jackson, Boostrom, & Hansen, 1993). Five, classified as *moral instruction*, are "avowedly moral" (p. 3) and, thus, relatively visible: (a) moral instruction as independent curricula; (b) deliberate moral instruction within academic curricula; (c) rituals and ceremonies engendering pride, loyalty, reverence, piety, and thankfulness; (d) visual displays with moral content; and (e) spontaneous moral commentary. Classified as *moral practice*, the remaining three categories reflect the personal qualities, understandings, beliefs, assumptions, and presuppositions of teachers and, as such, "*embody* the moral [italics in original]" (p. 3). They include the following: (a) classroom rules and regulations as manifestations of fairness, equity, and care; (b) curricular substructures that shape the order, organization, and delivery of curricula as manifestations of truthfulness, worthwhileness, fairness, and justice; and (c) expressive morality in the teachers' facial expressions, gestures, body posture, mannerisms, and style, and in the placement, quality, and quantity of furniture, objects, and display items in the learning environment as manifestations of care, respect, and trust.

Similarly, The Manner in Teaching Project (Fenstermacher, 2001; Richardson & Fenstermacher, 2000, 2001) identified six methods by which teachers attempt to foster moral conduct, improve intellectual dispositions, and cultivate virtue: (a) constructing classroom communities, characterized by mutual respect, sharing, tolerance, orderliness, and productive work; (b) didactic instruction that connotes desired moral and intellectual conduct; (c) design and execution of academic task-structures to cultivate intellectual virtues, such as critical thinking, regard for truth, and respect for evidence; (d) publicly calling out for conduct of a particular kind; (e) private conversations with students; and (f) showcasing specific students with exemplary behaviour so they may serve as role models for their peers. The investigators recognized that teachers also serve as role models for students but did not give modelling separate standing in this taxonomy, acknowledging the complex yet undetermined ways in which modelling is integrated with all six methods.

Character education, cognitive development and care ethics, the three main approaches to moral education, originally offered strategies particular to their own agendas, namely the cultivation of virtue, moral reasoning, and caring relationships, respectively. Yet, over time and with continued refinement, a wider range of strategies has emerged from each, with notable commonalities. Traditional character education primarily promoted direct, didactic instruction. Contemporary incarnations also address qualities of the learning environment, relationships between teachers and students, and teachers' conduct, behaviour, and practices, among other considerations. Ryan and Bohlin (1999), for example, recommend that teachers attend to six "*Es*", summarized as follows:

- *Example*: modelling morality and exemplary behaviour;
- *Explanation*: providing information and detail to enhance students' understandings;
- *Ethos* or *ethical environment*: creating a class community based on respect and fairness, among other ethical values;
- *Experience*: providing service-learning opportunities;
- *Exhortation*: inspiring and motivating students to be involved and do their best;
- *Expectations of excellence*: challenging students with high standards of achievement.

Lickona (2004) recommends the following seven practices:

- Building bonds and relationships with students and modelling character virtues;
- Teaching academics and character simultaneously and in support of each other;
- Practicing character-based discipline to reinforce accountability, responsibility, respect, and shared decision-making;
- Teaching manners, including appropriate greetings and responses;
- Preventing cruelty and promoting kindness among students;

- Helping students take responsibility for understanding, assessing, and building their own characters;
- Involving students in creating a school of character through leadership opportunities.

Lastly, as part of a 12-Point Comprehensive Approach Lickona, Davidson, and Khmelkov (Center for the 4th and 5th Rs, n.d.-a) propose nine classroom-based strategies for nurturing performance character and moral character:

- The teacher as caregiver, model, and mentor;
- An ethical learning community;
- Character-based discipline;
- A democratic classroom environment;
- Teaching character through curriculum;
- Cooperative learning;
- Conscience of craft;
- Ethical reflection;
- Teaching conflict resolution.

Among these three lists, several strategies are commonly articulated, including teacher modelling, attending to the learning environment, direct instruction, and providing relevant opportunities for student participation.

The traditional method for nurturing cognitive development and moral reasoning involves teachers facilitating discussions according to age-related developmental expectations. Nucci (2009) clarifies, "Effective discussions involve having students think about and reflect upon what they are hearing and offer suggestions that make use of what others have to say" (p. 110). The goal is not to generate a winner, as in debating, but to ensure that the best arguments and ideas emerge for the benefit of all. Topic sources include increasingly sophisticated and complex ethical conflicts and dilemmas, usually, but not only, hypothetical in nature, and moral issues embedded in academic curricula, particularly in the subject areas of social studies, mathematics, visual art, literature, and science (Howard, 2005; Nucci, 2001, 2008, 2009; Nucci & Turiel, 2009).

Social opportunities inherent in everyday classroom life and learning activities are also a means by which students' capacity for moral reasoning can be nurtured. According to Snarey and Samuelson (2008), moral cognitive abilities improve through interactions and experiences with others, particularly when children are encouraged to actively construct ways of thinking about right and wrong. Teachers might facilitate this by engaging students in developmental discipline practices (Nucci, 2009; Watson, 2008), a peer-mediated conflict resolution program (Howard, 2005; Johnson & Johnson, 2008), and service-learning activities (Billig, 2000, 2009; Howard, 2005). The Just Community Schools, developed by Lawrence Kohlberg and associates in the later part of the 20th century, is a more comprehensive approach. Students, teachers, and administrators participate in town hall–type meetings to collaboratively problem-solve and make decisions (Nucci, 2009; Power & Higgins-D'Alessandro, 2008). As stated by Nucci (2009), "Students are not empowered to change the basic academic framework of the school, but rather those norms that pertain to problematic areas of their moral interactions" (pp. 80–81). This might entail challenging the fairness of rules, behavioural guidelines, and select policies, for example (Oser, Althof, & Higgins-D'Alessandro, 2008).

Finally, care ethics theory alleges that reciprocally caring relationships between teachers and students naturally support the flourishing of several moral values, including care, love, kindness, compassion, empathy, sensitivity, trust, and respect and, thus, provide a context in which morality is both conveyed and practiced (Noddings, 2002, 2008, 2010). Character educators Lickona, Davidson, and Khmelkov agree: "Children need to form caring attachments to adults. These caring relationships will foster both the desire to learn and the desire to be a good person" (Center for the 4th and 5th Rs, n.d.-a). Nurturing such relationships, therefore, is understood as a strategy for moral education. Noddings (2008, 2010) further recommends that teachers utilize their influence in these relationships to teach morality by way of four components: (a) modelling moral behaviour and conduct to demonstrate what caring for others entails; (b) engaging with students in authentic dialogue to nurture and sustain the relationship; (c) providing opportunities for students to practice caring for others so they develop that capacity; and (d) confirming each child's goodness to attribute the best possible motives in the case of unacceptable behaviour. Although modelling and dialogue with students, in particular, are widely recommended, Noddings (2010) claims that caring relationships are the necessary context for these strategies to be effective as moral education: "In efforts at moral education, effective modeling requires a caring relation; for dialogue to be genuine, such a relation is essential" (p. 148).

To explore Terry's moral education strategies, I pose the question: What strategies does the teacher use to impart morality to students? Table 2.3 summarizes the discussion above and anticipates possible answers according to four sources: (a) the classroom environment; (b) the teacher's personal morality; (c) general pedagogy; and (d) direct instruction.

Classroom Environment

Sizer and Sizer (1999) contend, "People teach, but the institutions which people build also teach" (p. xiii). Dewey argued this concept a century earlier. As social institutions, he believed that schools expose children to the communal life of society and encourage social interaction. Thus, in an environment characterized by democratic values, students might cultivate moral knowledge, sound judgment, and responsibility and find it both possible and

Table 2.3   Strategies

Classroom environment:	• democracy, community, ethic of community • Laboratory Schools, Just Community Schools
physical set-up	• expressive morality (e.g., placement, quality, quantity of furniture, objects, display items) • access to supplies and materials • visual displays (e.g., posters, banners, signs)
classroom life	• culture (e.g., rituals, ceremonies, behavioural expectations) • structures (e.g., rules, regulations, duties, processes)
Teacher's personal morality:	• the teacher as a moral person
modelling	• expressive morality (e.g., actions, behaviours, conduct, facial expressions, gestures, body language, posture, mannerisms, style) • attitudes, beliefs • decision-making, problem-solving, conflict resolution • ongoing personal development • coincidental/unknowing, purposeful/intentional
relationships with students	• caring, warm, nurturing, trusting • reciprocal, as in care ethics
General pedagogy:	• professional practices
academic curricula	• curricular substructures • design and execution of academic tasks • lens of moral and existential questions
behaviour management	• aims for self-regulation, self-control, self-discipline • developmental discipline (e.g., social problem solving, positive feedback, judicious use of consequences) • character-based discipline (e.g., rules, consequences)
Direct instruction:	• overt moral education
preplanned lessons	• didactic instruction as independent curricula (e.g., program, unit of study) • deliberate didactic instruction within academic curricula (e.g., probing/guiding questions)
spontaneous messages	• spontaneous moral commentary • publicly (e.g., callouts, showcasing) • private conversation

desirable to be good. Dewey put this theory to the test in 1896, with the University of Chicago Laboratory Schools (ucls.uchicago.edu/index.aspx). Still in operation, the schools encourage students to collaborate with teachers and administrators in shaping their own learning and schooling experiences, while also learning to express citizenship and moral values (Dill,

2007; Hansen, 2002; Noddings, 1998). This notion of classroom democracy as a means of moral education was explored again in Kohlberg's Just Community Schools and is currently recommended by Lickona, Davidson, and Khmelkov in their 12-Point Comprehensive Approach to character education (Center for the 4th and 5th Rs, n.d.-a).

The metaphor of community is also used to define a school and classroom environment in which students feel a sense of connection, belonging, autonomy, competence, self-worth, caring, fairness, justice, respect, tolerance, and support, and within which students' moral growth and development is prioritized (Richardson & Fenstermacher, 2000, 2001; Solomon, Watson, Battistich, Schaps, & Delucchi, 1996). Ryan and Bohlin (1999) submit that a virtues-infused community helps students to know, respect, affirm, value, and care for each other. As one of their six "*Es*", they advise creating a class community based on moral values of respect and fairness. Lickona (2004) recommends developing a caring community to prevent peer cruelty and promote kindness. The Manner in Teaching Project identified the construction of class communities as a method for fostering moral and intellectual virtues (Fenstermacher, 2001). Fenstermacher, Osguthorpe, and Sanger (2009) argue that moral education programs are not fully effective unless embedded in morally attentive school and class communities. Noddings (2002) asserts that morality flourishes in a class community and that teachers should assume a moral obligation to create this learning environment. Finally, Furman (2004) concurs with this latter point in proposing an ethic of community to complement and extend Starratt's (1994) multidimensional ethical framework of justice, critique, and care. In outlining communal processes for achieving community, Furman echoes earlier work on democratic classrooms, explicitly the full participation of all community members.

Whether characterized as democracies or communities, or with particular moral values, classroom environments that are morally educative are created by attending to the physical features of the classroom's space and the cultural and structural characteristics of classroom life. Goodman and Balamore (2003) explain:

> One wants each student to feel a rush of gladness when entering the room that first day, to sense, "Here I am safe, here I can be myself, here everyone is treated fairly, here there are lots of possibilities". That means designing a room to look attractive, colourful, and enticing, yes, but also one that is "fair", where each child finds nourishment for his or her interests and aptitudes. . . . "In this room this year, we will develop a respectful, caring, lively, freedom-loving, hard-working, engaged community of learners". (p. 12)

The Moral Life of Schools Project (Jackson, Boostrom, & Hansen, 1993) addresses the room's physical set-up in two categories, *expressive morality* and *visual displays*. Regarding expressive morality, the placement, quality, and quantity of furniture, objects, and display items play a role in facilitating interactions with the teacher and among students and may convey care and sensitivity. Visual displays with moral content, such as posters, banners, and signs, indicate teachers' expectations for how one is to behave and associate with others. Similarly, The Manner in Teaching Project (Fenstermacher, 2001; Richardson & Fenstermacher, 2000, 2001) acknowledges classroom set-up in its method of *constructing classroom communities*. According to Fenstermacher (2001), "How the teacher sets up the furniture and arranges for student access to supplies and materials in the classroom also signals appropriate and inappropriate conduct" (p. 643).

Cultural and structural characteristics of classroom life that impart morality include rituals, ceremonies, rules, regulations, duties, behavioural expectations, and processes. In two categories of moral influence, The Moral Life of Schools Project (Jackson, Boostrom, & Hansen, 1993) illustrates that rituals and ceremonies may engender values of pride, loyalty, reverence, piety, and thankfulness, and rules and regulations may uphold fairness, equity, and care. The Manner in Teaching Project (Fenstermacher, 2001; Richardson & Fenstermacher, 2000, 2001) suggests that teachers' rules, duties, and behavioural expectations are means of creating a class community defined by mutual respect, sharing, tolerance, orderliness, and productive work. Finally, democratic processes, such as equal participation, collaboration, cooperation, and consultation, in decision-making, problem-solving and peer-mediated conflict resolution are believed to help students develop moral judgement, reflection, and rationality; promote expressions of several moral values; and cultivate student relationships (Oser, Althof, & Higgins-D'Alessandro, 2008; Power, 1988).

Teacher's Personal Morality

Ryan and Bohlin (1999) note, "Ultimately, it is the *person* [italics in original], not the teacher, who makes a lasting impression on his or her students" (p. 142). Who teachers are as people and what they think, believe, and assume, broadly and situationally, are made visible by their actions, behaviours, conduct, facial expressions, gestures, body language, posture, mannerisms, and style. The Moral Life of Schools Project (Jackson, Boostrom, & Hansen, 1993) calls this *expressive morality*. As a strategy for moral education, expressive morality is essentially modelling and, thus, pervades all aspects of teaching practice, and school and classroom life (Fenstermacher, 2001; Richardson & Fenstermacher, 2000, 2001; Jackson, Boostrom, & Hansen, 1993). For example, when teachers commit to helping students with personal problems, they model compassion, kindness, sensitivity, and empathy. When they admit an error and apologize, they model honesty, humility, and respect. When they confront colleagues regarding gossip, they model courage, confidence, and responsibility (Campbell, 2003; Ryan & Bohlin, 1999; Sockett, 1993; Wynne & Ryan, 1997). Hansen (1993) suggests that such modelling occurs naturally, coincidentally, and unknowingly as teachers carry out their professional responsibilities:

> This is not to suggest that [the teacher] is trying to "demonstrate" sensitivity. Rather, her actions can be seen as those of a person who *is* sensitive [quotations and italics in original], which means that the everyday enactment of that disposition in her work need not be tied to self-conscious intent. (p. 659)

Campbell (2003) accepts this claim but also empirically demonstrates that some teachers are, in fact, acutely aware of modelling morality for students and do so purposefully and intentionally by way of demonstration.

Modelling assumes that students will *catch* what is modelled by influential others. Wynne and Ryan (1997) express this in saying, "If these adults are respectful and caring, the children will try to be respectful and caring. If the adults are self-serving and lazy, the children will be self-serving and lazy" (p. 121). Therefore, a person with a good and righteous disposition is believed to affect the same in others (Fenstermacher, 2001). It is also assumed that modelling morality substantiates and authenticates other methods of moral teaching (Fenstermacher, 2001; Richardson & Fenstermacher, 2000, 2001). Fenstermacher, Osguthorpe, and Sanger (2009) argue, "Morality taught through content in the absence of moral manner on the part of the teacher will ring false to students and likely not be seriously entertained by them" (p. 11). These positions are intuitively accepted and generally agreed upon. Yet, Schwartz (2007) reminds us that the process of modelling is not well-understood: "Teachers need to be modelers of moral character but there is no clearly accepted definition of what moral character is, no accepted understanding for how it is modeled, and no generally accepted means for measuring it" (p. 1). Regardless, the expectation to be exemplars holds teachers to a high moral standard in their personal and professional lives. Goodman and Lesnick (2001) note, "To be credible to children, teachers must attend first to their own morality" (p. 271). This includes sustaining virtuous attitudes and beliefs; practicing ethical decision-making, problem-solving, and conflict resolution; and committing to one's ongoing moral development (Arthur, 2008; Campbell, 2003; Lickona, 1991, 2004; Noddings, 2008; Schwartz 2007).

Caring relationships between teachers and students are thought to enhance the morally instructive influence of teachers' personal morality (Noddings, 2010). This notion is reflected in the 12-Point Comprehensive Approach. The category called *teacher as caregiver, model and mentor* proposes:

> Values are best transmitted through these warm, caring relationships. In schools, as in families, kids care about our values because they know we care about them. If children do not experience an adult as someone who respects and cares about them, they are not likely to be open to anything the adult wishes to teach them about values. (Center for the 4th and 5th Rs, n.d.-a)

Accordingly, Ryan and Bohlin (1999) propose that teachers get to know students so well that they know even their implicit lives. Lickona and Davidson (2005) agree, warning, "When a teacher does not know a student well, it is easy for that student to cheat, cut corners, and 'fake it' [quotations in original]" (p. 34). They also claim that such relationships supersede academic affairs: "[Students] don't care how much you know until they know how much you care" (Lickona & Davidson, 2005, p. 90). Watson (2003, 2008) recommends relationships that are warm, nurturing, and trusting as a condition for motivating students to inherently do what is right and good. Sockett (1993) recommends relationships that are caring, fair, honest, and trusting. While teacher-student relationships are widely endorsed in moral education literature, care ethics (Noddings, 2002, 2008, 2010) has developed the most comprehensive articulation, noting that the reciprocal responsibility to nurture and preserve such relationships provides students with additional opportunities to practice expressing a range of moral values themselves.

Morally educative claims regarding teacher-student relationships are primarily from theoretical and practical accounts. Neither The Moral Life of Schools Project (Jackson, Boostrom, & Hansen, 1993) nor The Manner in Teaching Project (Fenstermacher, 2001; Richardson & Fenstermacher, 2000, 2001) explicitly included such relationships in their taxonomies. The Manner in Teaching Project makes passing reference to relationships in the *constructing classroom communities* method, but as an outcome of moral learning, rather than a context or strategy for teaching morality: "The teachers at Jordan Elementary are more likely to imply to their students that they need to behave in a particular way if they are to have a successful relationship to the teacher" (Fenstermacher, 2001, pp. 642–643). Further, The Moral and Ethical Bases of Teachers' Interactions with Students study (Campbell, 2003, 2004; Campbell & Thiessen, 2000, 2001) determined that moral agent teachers prioritize expressions of care in their practice, but as a virtue of character, not in the relational sense.

General Pedagogy

Moral education strategies are embedded in professional practices related to the delivery of academic curricula and the management of classroom behaviour. The *curricular substructures* category of The Moral Life of Schools Project (Jackson, Boostrom, & Hansen, 1993) and the *design and execution of academic task-structures* activity of The Manner in Teaching Project (Fenstermacher, 2001; Richardson & Fenstermacher, 2000, 2001) suggest that moral values are communicated by how teachers shape, order, organize, and present curricula and how they assess student learning. For example, when teachers present well-planned lessons that are enthusiastically taught, they convey diligence and care. When teachers involve all students in a class discussion, they convey inclusiveness and belonging. When teachers accommodate for different learning abilities, they convey respect and fairness. When teachers interpose lectures with activities, particularly for younger children, they convey thoughtfulness and sensitivity. When teachers administer a quiz to ensure homework has been completed, they convey responsibility and integrity, and when teachers assess and return student work promptly, they convey dependability and responsibility. The Moral and Ethical Bases of Teachers' Interactions with Students study (Campbell, 2003, 2004; Campbell & Thiessen, 2000, 2001) identified teachers' moral intentions regarding several such practices.

Recognizing embedded morality in existing academic curricula, Simon (2001) recommends curricular reorganization to formally integrate subject areas around both moral and existential questions. In distinguishing the two, she suggests that morality is concerned with how humans should act or should have acted in situations that involve the well-being of self, others, other living things, and the earth. Relevant exploratory questions might include: How can we respond to human suffering in ways that promote dignity? What, if anything, constitutes a just war? How should a society distribute its wealth? Are there scientific discoveries that humans should not pursue? Does democracy result in representative and humane governance? What is the impact of particular technological innovations on the environment? Although related, existentialism inquires into human nature, the mysteries of the universe, and the quality of our physical, spiritual, or emotional existence. Exploratory questions that are existential in nature include: What forces give rise to cruelty among human beings? What does it mean to be a *good* human being? What do I need to do to promote my own health and happiness? What gives my life meaning? How does human life differ from other kinds of life on earth? In accordance with this approach, Sanger and Osguthorpe (2005) describe how a teacher paired Mary Shelley's novel *Frankenstein* with popular scientific articles on genetically modified organisms and the myths of Prometheus and Pandora. Students were asked to extract, explore, and compare inherent moral conflicts. English, science, and social studies subject areas were integrated with the use of multiple resource materials.

How teachers manage classroom behaviour may also be morally educative. Moral education literature widely recommends an approach that encourages students' self-regulation, self-control, and self-discipline. This is the basic premise of both developmental discipline (Nucci, 2009; Watson, 2003, 2008) and character-based discipline (Lickona, 2004). Developmental discipline is consistent with cognitive development theories and, as such, assumes that children construct understandings of moral conduct and behaviour for themselves through intrinsic developmental processes that are stimulated and encouraged by adults. Nucci (2009) recommends methods of social problem-solving, positive feedback, and the judicious use of consequences. Character-based discipline is similarly instructive but arises from a different philosophical position. Character educators assume that "we are born both self-centered and ignorant, with our primitive impulses reigning over reason" (Ryan & Bohlin, 1999, pp. 5–6). Thus, character education, generally, and character-based discipline, specifically, provide children with what they naturally lack. Accordingly, Lickona (2004) contends, "[Discipline] has to help [students] develop the virtues—often respect, empathy, good judgment, and self-control—whose absence led to the discipline problem in the first place" (p. 144). The 12-Point Comprehensive Approach to Character Education (Center for the 4th and 5th Rs, n.d.-a) recommends the use of rules and consequences that students determine in collaboration with their teachers.

These examples of imparting morality by way of general pedagogy suggest that teachers do not need to reinvent their professional practices. Teaching is, on balance, a moral endeavour. Yet, I do not assume that good and right pedagogy, although essential, is necessarily a moral education strategy, particularly if the moral messages are not imparted with intent and remain hidden from students. The educative potential inherent in delivering academic curricula and managing classroom behaviour, therefore, is realized when the moral core is exposed.

Direct Instruction

Direct moral instruction represents the most overt expression of moral education, encompassing both preplanned lessons and spontaneous messages. The Moral Life of Schools Project (Jackson, Boostrom, & Hansen, 1993) and The Manner in Teaching Project (Fenstermacher, 2001; Richardson & Fenstermacher, 2000, 2001) demonstrated that teachers deliver preplanned, formal moral lessons to their students as an independent curriculum or as part of academic curricula. The Moral Life of Schools Project proposes two separate categories: *moral instruction as independent curricula* and *deliberate moral instruction within the academic curricula*. The Manner in Teaching Project proposes only a single method—*didactic instruction that directly signifies desired moral conduct*—but recognizes both curricular sources. Independent curricula might involve a character education program organized around selected moral values; a cognitive development program organized around ethical dilemmas; or smaller units of study. For example, McCadden (1998) developed a unit of study on friendship that involved telling stories, brainstorming, and discussing a variety of moral values he hoped to infuse into the culture of his research class. In a unit of study on heroes, Boone (2005) used narratives of moral role models to similarly impart morality to her research class.

Academic curricula also provide opportunities for teachers and students to engage in moral exploration. In the context of social studies and history, for example, one may examine struggles for justice; individuals whose actions, moral or otherwise, have changed the course of history; and concepts of personhood and society that highlight human diversity (Lickona, 2004; Straughan, 1988). Fictional literature is replete with characters embroiled in conflicts and dilemmas through which one may explore moral standards and models of goodness and evil (Bryan, 2005; Edgington,

2002). Scientific discoveries and technological developments raise ethical issues relating to responsible applications, the social and political contexts in which pioneering scientists live and work, and the use of animals for testing (Fenstermacher & Osguthorpe, 2000; Iozzi & Paradise-Maul, 1980). Physical education provides an opportunity to grapple with and practice competitive and cooperative values in sports (Halstead & Taylor, 2000). Mathematics can be used to assess the fairness of differential insurance rates based on age and gender, as well as issues related to justice and equality (Falkenberg & Noyes, 2010; Howard, 2005). Finally, the moral obligation of museums to return artwork stolen from the Jewish community of Eastern Europe during the Second World War can be debated in a visual arts course (Howard, 2005). The possibilities seem endless, even without having to realign or reorganize subject content, as Simon (2001) suggests.

Although this approach is broadly endorsed in moral education literature, Ryan (1993) warns, "Simply selecting the curriculum is not enough; like a vein of precious metal, the teacher and students must mine it together" (p. 17). Accordingly, Wynne and Ryan (1997) propose probing questions to guide student deliberations. These include, for example, why did the people believe they were acting morally, and what does this indicate about how we treat strangers? They also suggest that students might be asked to make an argument for the immorality of a situation, character, or decision. Such questions are more specific in nature than the existential and moral questions similarly proposed by Simon (2001) and recounted above.

Spontaneous moral messages are unplanned and informal but no less direct and intentional. In a broad category, called *spontaneous moral commentary*, The Moral Life of Schools Project (Jackson, Boostrom, & Hansen, 1993) illustrated that teachers seize naturally occurring opportunities to communicate messages grounded in moral values or contextualized as good and right. Such opportunities may arise from any aspect of teaching, learning, and classroom life, including academic work, pedagogical activities, class management, events, and incidents as teachers probe students' thinking for deeper meaning and understanding, uphold high expectations for student conduct and academic work, or deal with problems, decisions, and conflicts. For example, Sockett (1993) recalls a teacher's public request for students to stay clear of the freshly painted classroom door until it had dried to demonstrate respect and consideration for the painter, who had worked hard, and for their mothers, who would have to clean their clothes should the clothing become soiled with paint. Similarly, The Manner in Teaching Project (Fenstermacher, 2001; Richardson & Fenstermacher, 2000, 2001) identifies three such methods for spontaneously imparting moral messages, mostly in regard to student behaviour. *Calling out for conduct of a particular kind* and *showcasing specific students* are public exhortations. *Private conversations* convey moral messages solely to those involved.

These strategies have been presented positively as a means of enhancing the moral growth and development of students. Oser (1994) cautions, however, that every action designated to enhance moral development may entail a morally negative side effect. He identifies group work, in particular. This pedagogical strategy is often suggested as a means of facilitating relationships among students and teaching pro-social and moral values. Yet, Oser (1994) asserts:

> Students feel confused or even ashamed if the distribution of the different tasks within one group is unequal; they may feel humiliated or frustrated when classmates with higher capacities work on their own and, thus, do not share their experiences and thinking with those who are weaker in the given subject. Similarly, students may feel hurt when a number of friends work together in one group and leave others out (and thus label them "outsiders"). (p. 58)

Wynne and Ryan (1997) express similar concerns regarding cooperative learning programs. Discussions emerging from academic curricula may also trigger feelings of vulnerability and isolation. This might be the case for students of Japanese descent if discussing the internment of Japanese Canadians during the Second World War. A formal lesson on "Christian" virtues may alienate those who are not Christian, even when the virtues are universally recognized. Finally, in fostering relationships with students, the teacher likely will not connect as well with some students as with others. Although well-meaning in most cases, the strategies outlined above must be carefully and thoughtfully implemented and monitored to ensure desired learning outcomes and to avoid or mediate undesirable consequences, particularly those that might be considered unethical or immoral.

ASSESSING SOCIAL-MORAL PROGRESS

Hugh Hartshorne and Mark May's character education inquiry (1930) was not conclusive regarding the effectiveness of character education. Nonetheless, the study initiated the development of many instruments and tools for assessing students' knowledge, skills, attitudes, opinions, motivations, and behaviour related to honesty and deceit, altruism, and pro-social values (Leming, 2008). Today, instruments extend beyond character education to include other forms of moral education, and beyond student assessment to include assessment of the learning environment as a place where morality flourishes, and of teachers' efficacy as moral educators. The Character Education Partnership (CEP) (Character.org, n.d.) collects and collates many such instruments, according to three categories—individual assessments, teacher assessments, and school assessments. Individual assessments pertain to students' character development. They include student self-reports and teacher reports regarding students' attitudes, behaviours, and/or skills. Some may be used prior to a particular character program, unit, or curriculum, and then immediately following to determine the program's impact. Teacher assessments inform teachers of their own attitudes, social-emotional abilities, and competence as moral educators. They may be used to promote personal and professional development, to build self-awareness, and to monitor progress. School assessments measure organizational culture and climate and, as such, include all stakeholders. Although primarily focused at the school-level, this category also includes the Classroom Environment Scale, which considers the effect of teaching methods, teachers' personalities, course content, class composition, and classroom environment. Most of the assessment tools identified below can be accessed through this resource.

The Center for the 4th and 5th Rs (n.d.-b) has developed several assessments related to character education, ranging in scope to measure the effectiveness of a character program, identify character development practices and school-wide behaviours of respect and responsibility, evaluate the school as a caring community, determine coaching approaches to character development, and gauge character development outcomes in a sports setting. The Twelve Component Assessment and Planning Tool is arguably the most comprehensive. A companion piece for the 12-Point Comprehensive Approach to Character Education, this instrument provides outcome indicators for 12 components of schooling: the teacher as caregiver, model and mentor; an ethical learning community; character-based discipline; a democratic classroom environment; teaching character through curriculum; cooperative learning; conscience of craft; ethical reflection; teaching conflict resolution; creating a culture of excellence and ethics; caring beyond the classroom; and schools, parents, and communities as partners.

Cognitive development theory generates tools for assessing one's developing framework and sophistication in making moral judgements and decisions. Among the most well-known instruments are those derived from the work of moral psychologist James Rest. In 1979, Rest devised the Defining Issues Test (DIT) as an alternative to Kohlberg's interview method. Participants are provided with the following six ethical dilemmas: (a) Should Heinz steal a drug from an inventor in town to save his wife, who is dying and needs the drug? (b) Should a man who escaped from prison but has since been leading an exemplary life be reported to authorities? (c) Should a student newspaper be stopped by a principal of a high school when the newspaper stirs controversy in the community? (d) Should a doctor give an overdose of painkiller to a suffering patient? (e) Should a minority member be hired for a job when the community is biased? (f) Should students take over an administration building in protest of the Vietnam War? Each dilemma is accompanied by 12 statements, or standard items, which students rank in importance for processing the dilemma. This assessment is now referred to as DIT-1. The DIT–short form assessment consists of the first three ethical dilemmas, and DIT-2 offers five ethical dilemmas. Answer sheets are scored by the Center for the Study of Ethical Development, at The University of Alabama (n.d.).

Recognizing that relatively little is known about teacher self-efficacy in regard to moral education, Narváez, Vaydich, Turner, and Khmelkov (2008) created the Teacher Efficacy for Moral Education measure (TEME). This tool assesses teachers' beliefs in their efforts to bring about improvement in students' moral character and behaviour. Five scales are used. Two scales relate to instruction (efficacy for promoting positive relationships and efficacy for helping students learn); one scale relates to teachers' self-efficacy for promoting character education; and two scales relate to school climate (the school culture scale and a collective efficacy measure). Each includes several statements, such as: "If students in my class seem discouraged about learning, I know how to get them feeling positively again"; "I am usually comfortable discussing issues of right and wrong with my students"; "Teachers who encourage responsibility at school can influence students' level of responsibility outside of school"; "Students and teachers trust each other"; and "Students help each other even if they are not friends" (Narváez et al., 2008, p. 6). Participants rate the statements using a 1 to 5 Likert-type system (1 = not at all true; 5 = very true).

Also on the topic of teacher assessment, Schwartz (2007) developed the Seven-Item Attribute Questionnaire (SIAQ) for evaluating teachers' modelled moral character. Seven attributes are organized across three perspectives—cognitive, affective, and action. The cognitive perspective contains a single attribute, paraphrased as the demonstration of self-reflection and reasoning skill. Four affective attributes include the following: demonstration of empathy and perspective-taking; obvious moral concern and care for others; leeway granted to self and others; and regulation of the teacher's own behaviour and emotions in accordance with the social good of others. Two attributes relate to the action perspective: congruence between the teacher's moral statements, understandings, and actions, and engaging in actions that indicate a commitment to the intellectual and/or emotional development of others. Teachers complete a version of the questionnaire pertaining to themselves, and students complete another version pertaining to their teachers. For example, item five on the teachers' version reads: "You share with students how you think through problems and admit when you make a mistake" (p. 25). The corresponding item on the students' versions reads: "My teacher shares with students how he or she thinks through problems and admits when he or she makes a mistake" (p. 23). A rating scale of 1 to 5 is used, 1 indicating "too little or no extent", and 5 indicating "to a very great extent" (p. 22).

These examples represent a larger pool of assessments that are primarily instrument-driven, objectively applied, quantitatively measured, and scientifically validated. Yet, Romanowski (2005) observes, "It is nearly impossible to measure the results of a [character education program] in a quantitative form because character education is a long term plan that materializes in students' adult lives" (p. 18). In accordance with the paradigm shift in education research, he recommends a qualitative approach to assessment. This coincides with Campbell's (2003) empirical work, which reveals that some teachers reflect on and evaluate the success of the moral

education they impart in relation to how they personally envision their classrooms and students. Hammerness (2006) concurs, saying, "Like a mirror, teachers compare daily practice to their vision and recognize successes as well as identifying areas for improvement" (p. 3). In addition, Nucci (2001) criticizes independent program-based assessments, believing that moral and social reasoning are best evaluated in the context of academic curricula. By example, he cites students' responses on schoolwork as having the potential to demonstrate increased moral reciprocity and decreased egocentricity.

To explore how Terry assesses her students' moral learning, I pose the question: How does the teacher assess the effectiveness of the moral education he or she imparts? Table 2.4 summarizes the discussion above, anticipating possible answers according to the following: (a) summative student assessment, (b) formative student assessment, (c) assessment of the classroom

Table 2.4   Assessment

Summative:	• culminating (e.g., quizzes, tests, exams, assignments, questionnaires, interviews, case study reports, research projects, essays, responsive writing, oral presentations, recitals, skits, role-playing) • commercial tools (e.g., CEP, Center for the 4th and 5th Rs, DIT-1, DIT-2, DIT–short form) • according to standards and benchmarks
Formative:	• ongoing (e.g., observation, conferences, interviews, feedback, self-assessment, socio-dynamic map, concept maps, study notes, thesis statements, essay outlines) • according to intuition, instinct, experience, expectations, perception, reflection, subjective judgment
Classroom culture and climate:	• track norms of behaviour; shared values, assumptions, beliefs; collaboration in problem solving, conflict resolution, decision-making; relationships • in accordance with expectations, vision, intuition, instinct, experience, perception, reflection, subjective judgment • commercial tools (e.g., CEP, Center for the 4th and 5th Rs, TEME)
Teacher self-assessment:	• commercial tools (e.g. TEME, SIAQ, CEP, Center for the 4th and 5th Rs, administrator evaluations, student assessments) • classroom culture/climate • feedback from students, parents, administrators, colleagues

culture and climate, and (d) teacher self-assessment. Summative assessments entail formal cumulative and culminating activities that are conducted at the completion of a topic, unit, semester, or year. Prior learning is usually measured against a predetermined standard or benchmark. Examples include quizzes, tests, exams, assignments, questionnaires, interviews, case study reports, research projects, essays, responsive writing, oral presentations, recitals, skits, or improvisational role-playing. Many of the assessment tools noted above can be used in this way. Formative assessment is ongoing and generally less formal. It takes place throughout a topic, unit, semester, or year, allowing teachers to monitor student progress as learning occurs. This might involve observing students at work and play; student-teacher conferences and interviews; student feedback and self-assessment; socio-dynamic mappings of class relationships; or a review of work in progress, including concept maps, study notes, thesis statements, and essay outlines. With less emphasis on standards and benchmarks, formative assessment relies more heavily on teachers' intuition, instinct, experience, expectations, perception, reflection, and subjective judgment.

The classroom culture or climate may be assessed for how well it enables morality to flourish, as well as how it reflects students' developing moral orientations. The former recalls the previous discussion on classroom environment as a moral education strategy. Its moral nature, therefore, indirectly indicates the teacher's efficacy as a moral educator. The latter is a means of assessing students' moral learning, growth, and development. Norms of behaviour; shared values, assumptions, and beliefs; collaboration in problem solving, conflict resolution, and decision-making; and student relationships, for example, can be tracked by the teacher in relation to students' moral understandings, intentions, and efforts, and in accordance with the teacher's expectations and vision. This likely involves intuition, instinct, experience, perception, reflection, and subjective judgment. Relevant assessment tools are also available from CEP (Character.org, n.d.) and the Center for the 4th and 5th Rs (n.d.-b), and as part of the TEME measure (Narváez, Vaydich, Turner, & Khmelkov, 2008).

Lastly, several commercial tools are available to assess teachers' efficacy as moral educators. TEME (Narváez, Vaydich, Turner, & Khmelkov, 2008) and SIAQ (Schwartz, 2007) were outlined above. Other instruments are provided by CEP (Character.org, n.d.) and the Center for the 4th and 5th Rs (n.d.-b). These include the 8 Strengths Assessment, Coach's Character Development Self-Evaluation Checklist, Attitudes Toward Inclusive Education Scale, Emotional Quotient Inventory, Behavioral Characteristics of a Teacher, and Positive Teacher-Student Relations. Teachers might also adapt formal evaluations designed for administrators and utilize student assessments as an indication of their own success. Although a causal connection between the moral education teachers provide and students' moral learning, growth, and development is unclear, one is, nonetheless, assumed. Finally, teachers may solicit formal or informal feedback from students, parents, administrators, and colleagues.

OPPORTUNITIES

This framework represents a compilation of key theoretical, philosophical, practical, and empirical work related to moral education, the moral dimensions of teaching, learning, and classroom life, and moral agency. It is not an exhaustive exploration of these topics and themes but instead intends to offer a means of cohesion for an extensive body of scholarship that has been deemed difficult to synthesize, comprehend, and apply (Sanger, 2003; Sanger & Osguthorpe, 2005). By repositioning the literature according to four domains of teaching practice—vision, content, strategies, and assessment—rather than philosophy and theory, prevailing conceptual silos are disrupted. This is particularly significant in regard to character education, cognitive development, and care ethics. With dissimilar beliefs and assumptions, these three orientations to moral education are usually presented to educators as alternatives. An applied framework, however, enables consideration and exploration of a more inclusive, integrated, and harmonious approach in the classroom.

This framework also reveals the literature's limitations. The moral and ethical objectives of teachers' *vision* and the means by which teachers *assess* themselves as moral educators and their students' social-moral growth and development are less well represented than the moral *content* that is imparted and the moral education *strategies* that are used. Further, literature relating to the latter two domains, while extensive, is largely conceptual, theoretical, and philosophical. Regarding strategies, The Moral Life of Schools Project (Jackson, Boostrom, & Hansen, 1993) and The Manner in Teaching Project (Fenstermacher, 2001; Richardson & Fenstermacher, 2000, 2001) each provides a taxonomy of sorts, with some indication of knowledge and abilities content. These results are supported and extended by other studies (Buzzelli & Johnston, 2002; Campbell, 2003, 2004; Campbell & Thiessen, 2000, 2001; Fallona, 2000; McCadden, 1998; Richardson & Fallona, 2001; Simon, 2001). Yet, collectively this group of studies represents the practices of fewer than 100 teachers. Finally, The Moral and Ethical Bases of Teachers' Interactions with Students study (Campbell, 2003, 2004; Campbell & Thiessen, 2000, 2001) identified moral values which moral agent teachers prioritize in their professional practice, but not necessarily for the purpose of morally educating students.

In the chapters that follow, I take on these opportunities, exploring a more integrated approach to moral education and expanding insights and knowledge regarding teachers' moral agency. Chapter Three situates this profile within the physical and cultural environment of Middlevale School and within Terry's personal ideological orientation. These outer and inner contexts shape and enable seven key practices. Chapters Four and Five detail these practices from two positions. Chapter Four outlines three practices by which Terry teaches morally or in ways that are morally good and right. Chapter Five outlines four practices by which Terry, more directly, teaches morality. Collectively, these three chapters illustrate Terry's vision for classroom life and students' growth, development, and well-being; the moral knowledge and abilities Terry conveys to her students; the strategies by which she does so; and how Terry assesses her students' moral growth and development. The concluding chapter returns to the core concept of moral agency and suggests how this exemplary case study might contribute to teaching practice, teacher education, and academic research.

REFERENCES

Arthur, J. (2008). Traditional approaches to character education in Britain and America. In L. P. Nucci & D. Narváez (Eds.), *Handbook of moral and character education* (pp. 80–98). New York: Routledge.

Balch, M. F., Saller, K., & Szolomicki, S. (1993). Values education in American public schools: Have we come full circle? *Association for Supervision and Curriculum Development, 94,* 34. Retrieved March 25, 2009, from http://www.eric.ed.gov/ERICDocs/data/ericdocs2sql/content_storage_01/0000019b/80/15/94/95.pdf.

Beachum, F. D., & McCray, C. R. (2005, Summer). Changes and transformations in the philosophy of character education in the 20th century. *Essays in Education, 14,* 1–7. Retrieved February 25, 2010, from http://www.usca.edu/essays/vol142005/beachum.pdf.

Billig, S. H. (2000). The effects of service learning. *School Administrator, 57*(7), 14–18. Retrieved May 15, 2012, from http://proquest.com.

Billig, S. H. (2009). It's their serve. *Leadership for Student Activities, 37*(8), 8–13. Retrieved May 15, 2012, from http://proquest.com.

Boone, M. (2005). *Experiential storytelling as curriculum in elementary schools: A narrative approach.* (Doctoral dissertation, University of Toronto, Toronto).

Borba, M. (2001). *Building moral intelligence: The seven essential virtues that teach kids to do the right thing.* San Francisco: Jossey-Bass.

Bryan, L. (2005). Once upon a time: A Grimm approach to character education. *Journal of Social Studies Research, 29*(1), 3–6. Retrieved February 25, 2010, from http://proquest.umi.com.

Buzzelli, C. A., & Johnston, B. (2002). *The moral dimensions of teaching: Language, power, and culture in classroom interaction.* New York: RoutledgeFalmer.

Campbell, E. (2003). *The ethical teacher.* Maidenhead: Open University Press McGraw-Hill.

Campbell, E. (2004). Ethical bases of moral agency in teaching. *Teachers and Teaching: Theory and Practice, 10*(4), 409–428.

Campbell, E. (2008a). Teaching ethically as a moral condition of professionalism. In L. P. Nucci & D. Narváez (Eds.), *Handbook of moral and character education* (pp. 601–617). New York: Routledge.

Campbell, E. (2008b). The ethics of teaching as a moral profession. *Curriculum Inquiry, 38*(4), 357–385.

Campbell, E., & Thiessen, D. (2000, April). *Moral and ethical exchanges in classrooms: Preliminary findings.* Paper presented at the Annual Meeting of the American Educational Research Association, New Orleans, LA.

Campbell, E., & Thiessen, D. (2001, April). *Perspectives on the ethical bases of moral agency in teaching.* Paper presented at the Annual Meeting of the American Educational Research Association, Seattle, WA.

Center for the 4th and 5th Rs (n.d.-a). Our 12-Point Approach. Retrieved May 9, 2014, from http://www2.cortland.edu/centers/character/.

Center for the 4th and 5th Rs (n.d.-b). Assessment Instruments. Retrieved May 9, 2014, from http://www2.cortland.edu/centers/character/.

Center for the Study of Ethical Development, at The University of Alabama. (n.d.). The Center. Retrieved May 9, 2014, from http://ethicaldevelopment. ua.edu/.

Character.org. (n.d.). Assessment Tools. Retrieved May 9, 2014, from http://www. character.org/.

Cohen, L., & Manion, L. (1994). Introduction: The nature of inquiry. In L. Cohen & L. Manion (Eds.), *Research methods in education* (4th ed., pp. 1–43). London: Routledge.

Cohn, M. M., & Kottkamp, R. B. (1993). *Teachers: The missing voice in education.* Albany: State University of New York Press.

Coloroso, B. (2005). *Just because it's not wrong doesn't make it right: From toddlers to teens, teaching kids to think and act ethically.* Toronto: Viking Canada.

Cooper, B. (2010). In search of profound empathy in learning relationships: Understanding the mathematics of moral learning environments. *Journal of Moral Education, 39*(1), 79–99. doi: 10.1080/03057240903528717.

Denzin, N. K., & Lincoln, Y. S. (2000). The discipline and practice of qualitative research. In N. Denzin & Y. Lincoln (Eds.), *Handbook of qualitative research* (2nd ed., pp. 1–28). Thousand Oaks, CA: Sage.

Dewey, J. (1929). *The Sources of a Science of Education.* New York: Horace Liveright. Retrieved July 18, 2014, from https://archive.org/details/sourcesofascienc 009452mbp.

Dill, J. S. (2007). Durkheim and Dewey and the challenge of contemporary moral education. *Journal of Moral Education, 36*(2), 221–237. doi: 10.1080/0305724 0701325357.

Edgington, W. D. (2002, May/June). To promote character education, use literature for children and adolescents. *The Social Studies, 93*(3), 113–116. Retrieved February 25, 2010, from http://proquest.umi.com.

Falkenberg, T., & Noyes, A. (2010). Conditions for linking school mathematics and moral education: A case study. *Teaching and Teacher Education, 26,* 949–956.

Fallona, C. (2000). Manner in teaching: A study in observing and interpreting teachers' moral virtues. *Teaching and Teacher Education, 16,* 681–695.

Fenstermacher, G. D. (2001). On the concept of manner and its visibility in teaching practice. *Journal of curriculum studies, 33*(6), 639–653. doi: 10.1080/00220 270110049886.

Fenstermacher, G. D., & Osguthorpe, R. (2000, April). *The manner of teachers and the character of students: What distinguishes character education from The Manner Project?* Paper presented at the Annual Meeting of the American Educational Research Association, New Orleans, LA. Retrieved July 3, 2008, from http:// www-personal.umich.edu/~gfenster/.

Fenstermacher, G. D., Osguthorpe, R. D., & Sanger, M. N. (2009, Summer). Teaching morally and teaching morality. *Teacher Education Quarterly, 36*(3), 7–19. Retrieved February 25, 2010, from http://vnweb.hwwilsonweb.com.

Field, S. L. (1996, Winter). Historical Perspective on Character Education. *The Educational Forum, 60*(2), 118–123. doi: 10.1080/00131729609335112.

Fraenkel, J. R., & Wallen, N. E. (2003). *How to design and evaluate research in education* (5th ed.). New York: McGraw-Hill.

Furman, G. C. (2004). The ethic of community. *Journal of Educational Administration, 42*(2), 215–235.

Gilligan, C. (1982). *In a different voice: Psychological theory and women's development.* Cambridge, MA: Harvard University Press.

Giraffe Heroes Project. (n.d.). Home page. Retrieved March 10, 2014, from http://www.giraffe.org.

Glaze, A.E., Zegarac, G., & Giroux, D. (2008). *Finding Common Ground: Character Development in Ontario Schools, K-12.* Toronto, Ontario: Ministry of Education. Retrieved October 25, 2009, from http://www.edu.gov.on.ca.

Goodcharacter.com. (n.d.). Retrieved March 10, 2014, from http://www.goodcharacter.com.

Goodlad, J. I., Soder, R., & Sirotnik, K. A. (Eds.). (1990). *The moral dimensions of teaching.* San Francisco: Jossey-Bass.

Goodman, J. F., & Balamore, U. (2003). *Teaching goodness: Engaging the moral and academic promise of young children.* Boston: Pearson Education.

Goodman, J. F., & Lesnick, H. (2001). *The moral stake in education: Contested premises and practices.* New York: Addison Wesley Longman.

Greene, J. D. (n.d.). Retrieved April 2014, from http://www.wjh.harvard.edu/~jgreene/.

Halstead, J. M., & Taylor, M. J. (2000). Learning and teaching about values: A review of recent research. *Cambridge Journal of Education, 30*(2), 169–202. Retrieved February 15, 2009, from http://proquest.umi.com.

Hammerness, K. (1999, April). *Visions of delight, visions of doubt: The relationship between emotion and cognition in teachers' vision.* Paper presented at the Annual Meeting of the American Educational Research Association, Montreal, Quebec. Retrieved August 14, 2010, from http://ezproxy.qa.proquest.com/docview/62182843?accountid=14771.

Hammerness, K. (2002). *Learning to hope, or hoping to learn? The role of vision in the early professional lives of teachers.* Paper presented at the annual meeting of the American Educational Research Association, New Orleans, LA, in April 2000. Retrieved August 14, 2010, from http://ezproxy.qa.proquest.com/docview/62182843?accountid=14771.

Hammerness, K. (2006). *Seeing through teachers' eyes.* New York: Teachers College Press.

Hansen, D. T. (1993). From role to person: The moral layeredness of classroom teaching. *American Educational Research Journal, 30*(4), 651–674. doi: 10.3102/00028312030004651.

Hansen, D. T. (2002). The moral environment in an inner-city boys' high school. *Teaching and Teacher Education, 18*(2), 183–204.

Hartshorne, H., & May, M. A. (1930). A summary of the work of the character education inquiry. *Religious Education, 25,* 754–762. doi: 10.1080/0034408300250810.

Howard, R. W. (2005). Preparing moral educators in an era of standards-based reform. *Teacher Education Quarterly, 32*(4), 43–58. Retrieved February 26, 2010, from http://proquest.umi.com.

Howard, R. W., Berkowitz, M. W., & Schaeffer, E. F. (2004). Politics of character education. *Educational Policy, 18*(1), 188–215. doi: 10.1177/0895904803260031.

Howe, K., & Eisenhart, M. (1990). Standards for qualitative (and quantitative) research: A prolegomenon. *Educational Researcher, 19*(4), 2–9.

Hunt, T. C., & Mullins, M. (2005). *Moral education in America's schools: The continuing challenge.* Greenwich, CT: Information Age Publishing.

Iozzi, L. A., & Paradise-Maul, J. (1980). Issues at the interface of science, technology and society. In L. Kuhmerker, M. Mentkowski, & V. L. Erickson (Eds.), *Evaluating moral development and evaluating educational programs that have a value dimension* (pp. 131–138). Schenectady, NY: Character Research Press.

Jackson, P., Boostrom, R., & Hansen D. (1993). *The moral life of schools.* San Francisco: Jossey-Bass.

Johnson, D. W., & Johnson, R. T. (2008). Social interdependence, moral character and moral education. In L. P. Nucci & D. Narváez (Eds.), *Handbook of moral and character education* (pp. 204–229). New York: Routledge.

Joseph, P. B., & Efron, S. (2005, March). Seven worlds of moral education. *Phi Delta Kappan,* 525–533.

Josephson Institute. (n.d.). Character counts! Retrieved March 10, 2014, from http://charactercounts.org/.

Keewatin-Patricia District School Board. (n.d.) Retrieved October 25, 2009, from http://www.kpdsb.on.ca/default.aspx.

Kosnik, C. M., & Beck, C. (2009). *Priorities in teacher education: The 7 key elements of pre-service preparation.* London: Routledge.

Kosnik, C., Beck, C., Cleovoulou, Y., & Fletcher, T. (2009). Improving teacher education through longitudinal research: How studying our graduates led us to give priority to program planning and vision for teaching. *Studying Teacher Education, 5*(2), 163–175. doi: 10.1080/17425960903306880.

Krauss, S. E. (2005, December). Research paradigms and meaning making: A primer. *The Qualitative Report, 10*(4), 758–770. Retrieved March 12, 2008, from http://www.nova.edu/ssss/QR/QR10-4/krauss.pdf.

Lapsley, D. K. (2008). Moral self-identity as the aim of education. In L. P. Nucci & D. Narváez (Eds.), *Handbook of moral and character education* (pp. 30–52). New York: Routledge.

Leming, J. S. (2008). Research and practice in moral and character education: Loosely coupled phenomena. In L. P. Nucci & D. Narváez (Eds.), *Handbook of moral and character education* (pp. 134–157). New York: Routledge.

Leming, J. S. (2010). When research meets practice in values education: Lessons from the American experience. In T. Lovat, R. Toomey, & N. Clement (Eds.), *International research handbook on values education and student wellbeing* (pp. 91–110). New York: Springer Science + Business Media.

Lickona, T. (1991). *Education for character: How our schools can teach respect and responsibility.* New York: Bantam Books.

Lickona, T. (2004). *Character matters: How to help our children develop good judgement, integrity, and other essential virtues.* New York: Touchstone.

Lickona, T., & Davidson, M. (2005). *Smart & good high schools: Integrating excellence and ethics for success in school, work, and beyond.* Cortland, NY, Washington, DC: Center for the 4th and 5th Rs/Character Education Partnership. Retrieved January 3, 2009, from http://www.cortland.edu/character/.

McCadden, B. M. (1998). *It's hard to be good: Moral complexity, construction, and connection in a kindergarten classroom.* New York: Peter Lang Publishing.

McClellan, B. E. (1999). *Moral education in America: Schools and the shaping of character from colonial times to the present.* New York: Teachers College Press.

Narváez, D., & Lapsley, D. K. (2008, April–June). Teaching moral character: Two alternatives for teacher education. *The Teacher Educator, 43*(2), 156–172. doi: 10.1080/08878730701838983.

Narváez, D., Vaydich, J. L., Turner, J. C., & Khmelkov, V. (2008, July). Teacher self-efficacy for moral education: Measuring teacher self-efficacy for moral education. *Journal of Research in Character Education, 6*(2), 3–15.

Noddings, N. (1998, May). Thoughts on John Dewey's "Ethical principles underlying education". *The Elementary School Journal, 98*(5), 479–488. Retrieved February 10, 2010, from http://www.jstor.org/pss/1002326.

Noddings, N. (2002). *Educating moral people: A caring alternative to character education.* New York: Teachers College Press.

Noddings, N. (2008). Caring and moral education. In L. P. Nucci & D. Narváez (Eds.), *Handbook of moral and character education* (pp. 161–174). New York: Routledge.

Noddings, N. (2010). Moral education and caring. *Theory and Research in Education, 8*(2), 145–151. doi: 10.1177/1477878510368617.

Nucci, L. P. (2001). *Education in the moral domain.* New York: Cambridge University Press.

Nucci, L. P. (2008). Social cognitive domain theory and moral education. In L. P. Nucci & D. Narváez (Eds.), *Handbook of moral and character education* (pp. 291–309). New York: Routledge.

Nucci, L. P. (2009). *Nice is not enough: Facilitating moral development.* Upper Saddle River, NJ: Pearson Education.

Nucci, L., & Turiel, E. (2009). Capturing the complexity of moral development and education. *Mind, Brain, and Education, 3*(3), 151–159. doi: 10.1111/j.1751-228X.2009.01065.x.

Oser, F. (1994). Moral perspectives on teaching. In L. Darling-Hammond (Ed.), *Review of research in education* (pp. 57–127). Washington, DC: American Educational Research Association. Retrieved April 16, 2008, from http://www.jstor.org/stable/1167382.

Oser, F. K., Althof, W., & Higgins-D'Alessandro, A. (2008). The Just Community approach to moral education: system change or individual change? *Journal of Moral Education, 37*(3), 395–415. doi: 10.1080/03057240802227551.

Power, C. (1988). The just community approach to moral education. *Journal of Moral Education, 17*(3), 195–208. doi: 10.1080/0305724880170304.

Power, F. C., & Higgins-D'Alessandro, A. (2008). The just community approach to moral education and the moral atmosphere of the school. In L. P. Nucci & D. Narváez (Eds.), *Handbook of moral and character education* (pp. 230–247). New York: Routledge.

Rainbow District School Board. (n.d.). Character Education. Retrieved October 25, 2009. from http://www.rainbowschools.ca/programs/CharacterEducation/characterEducation.php.

Rest, J. R. (1986). *Moral development: Advances in research and theory.* New York: Praeger.

Richardson, V., & Fallona, C. (2001). Classroom management as method and manner. *The Journal of Curriculum Studies, 33*(6), 705–728. doi: 10.1080/00220270110053368.

Richardson, V., & Fenstermacher, G. D. (2000). *The Manner in Teaching Project.* Retrieved July 29, 2008, from www.personal.umich.edu/~gfenster.

Richardson, V., & Fenstermacher, G. D. (2001). Manner in teaching: The study in four parts. *The Journal of Curriculum Studies, 33*(6), 631–637.

Romanowski, M. H. (2005). Through the eyes of teachers: High school teachers' experiences with character education. *American Secondary Education, 34*(1), 6–23. Retrieved January 12, 2010, from http://proquest.umi.com.

Rosenberg, G. (2008). *Teachers' as moral agents to their students.* Unpublished manuscript, University of Toronto.

Ryan, K. (1988). Teacher education and moral education. *Journal of Teacher Education, 39,* 18–23. doi: 10.1177/002248718803900505.

Ryan, K. (1993) Mining the values in the curriculum. *Educational Leadership, 51*(3), 16–18. Retrieved April 16, 2011, from http://ezproxy.qa.proquest.com/docview/62786313?accountid=14771.

Ryan, K., & Bohlin, K. E. (1999). *Building character in schools: Practical ways to bring moral instruction to life.* San Francisco: Jossey-Bass.

Sanger, M. G. (2001). Talking to teachers and looking at practice in understanding the moral dimensions of teaching. *The Journal of Curriculum Studies, 33*(6), 683–704. doi: 10.1080/00220270110066733.

Sanger, M. N. (2003). *What is moral about teaching? A philosophical inquiry into the morally salient features of teaching.* (Doctoral dissertation, University of Michigan, Ann Arbor). Retrieved October 14, 2008, from http://proquest.umi.com.

Sanger, M. G., & Fenstermacher, G. D. (2000, April). *Aristotle is great—but is he enough? Expanding the theoretical grounds for inquiries into the moral dimensions of teaching.* Paper presented at the Annual Meeting of the American Educational Research Association, New Orleans, LA. Retrieved July 3, 2008, from http://www-personal.umich.edu/~gfenster/.

Sanger, M., & Osguthorpe, R. (2005). Making sense of approaches to moral education. *Journal of Moral Education, 34*(1), 57–71. doi: 10.1080/030572405000 49323.

Schuitema, J., ten Dam, G., & Veugelers, W. (2008). Teaching strategies for moral education: A review. *Journal of Curriculum Studies, 40*(1), 69–89. doi: 10.1080/00220270701294210.

Schwartz, M. J. (2007) The modeling of moral character for teachers: Behaviors, characteristics, and dispositions that may be taught. *Journal of Research in Character Education, 5*(1), 1–28. Retrieved December 11, 2009, from http://proquest.umi.com.

Simon, K. G. (2001) *Moral questions in the classroom: How to get kids to think deeply about real life and their schoolwork.* London: Yale University Press.

Sizer, T. R., & Sizer, N. F. (1999). *The students are watching: Schools and the moral contract.* Boston: Beacon Press.

Snarey, J., & Samuelson, P. (2008). Moral education in the cognitive developmental tradition: Lawrence Kohlberg's revolutionary ideas. In L. P. Nucci & D. Narváez (Eds.), *Handbook of moral and character education* (pp. 53–79). New York: Routledge.

Sockett, H. (1993). *The moral base for teacher professionalism.* New York: Teachers College Press.

Sockett, H. (2012). *Knowledge and virtue in teaching and learning: The primacy of dispositions.* New York: Routledge.

Solomon, D., Watson, M., Battistich, V., Schaps, E., & Delucchi, K. (1996, December). Creating classrooms that students experience as communities. *American Journal of Community Psychology, 24*(6), 719–748. Retrieved November 38, 2009, from http://www.metapress.com.

Starratt, R. J. (1994). *Building an ethical school: A practical response to the moral crisis in schools.* London: Falmer Press.

Stengel, B. S., & Tom, A. R. (2006). *Moral matters: Five ways to develop the moral life of schools.* New York: Teachers College Press.

Straughan, R. (1988). *Can we teach children to be good?* Philadelphia: Open University Press.

Watson, M. (2003). *Learning to trust: Transforming difficult elementary classrooms through developmental discipline.* San Francisco: Jossey-Bass.

Watson, M. (2008). Developmental discipline and moral education. In L. P. Nucci & D. Narváez (Eds.), *Handbook of moral and character education* (pp. 175–203). New York: Routledge.

Wynne, E. A., & Ryan, K. (1997). *Reclaiming our schools: Teaching character, academics, and discipline.* Upper Saddle River, NJ: Prentice-Hall.

York Region District School Board. (n.d.). Questions and Answers. Retrieved October 25, 2009, from http://www.yrdsb.edu.on.ca/.

3   Situating Terry's Practices
School Context and Personal Ideology

Terry's approach to moral education is enabled and supported by her school context and is shaped and motivated by her personal ideology. This chapter begins the portrait of moral agency by situating Terry's practices in these outer and inner worlds. The topics discussed include the school's physical features, structures, programs, and activities; Terry's classroom layout and class demographics; and Terry's personal background, philosophy of teaching, and vision for classroom life and students' growth, development, and well-being. Terry's voice sustains a strong presence as she recounts her own story and begins to share her innermost beliefs, assumptions, and perceptions.

THE CAMPUS

Inconspicuously nestled in the heart of an urban centre in Ontario, is a cluster of buildings, some over 100 years old. This constitutes the quirky campus of Middlevale School. Independent, nonresidential, and nonreligious, Middlevale provides comprehensive, coeducational programing that spans junior kindergarten to grade-eight. Approximately 350 students are organized among the buildings according to grade level and activity: junior kindergarten to grade-one in the primary building; grades two to five in the elementary building; grade-six in a third building; and grades seven and eight in the senior building. Some of these buildings are more thoughtfully named to honour local champions of education and human rights. These names are withheld here to protect the identity of the school.

Each division has a dedicated group of teachers for homeroom, language arts, science, social studies, health, and math. Art, music, drama, French, technology, and physical education are taught by specialty teachers who cross divisional lines and remain connected with students for many years. Unlike divisional teachers, specialty teachers and their programs are scattered among the various buildings. Students run through the courtyard, between buildings, in all weather to attend classes in the drama and art studios, French cabane, music rooms, library, computer and science labs, and gymnasium. These rooms are fully outfitted for their respective activities. For example, the drama studio is in the basement of the elementary building. Without windows, complete darkness is achievable. There is a stage with a curtain that can be drawn, a sound and lights booth, and a props and costumes cupboard. The gymnasium, on the main floor of the elementary building, has basketball nets, sports demarcation lines on the floor, and an electronic scoreboard. The primary music room, in the primary building, contains percussion instruments from different parts of the world. The senior music room, down the hallway from the drama studio, has traditional band instruments. Both have a piano. The art studio is in the basement of the primary building and is fully equipped for a variety of creative projects including painting, ceramics, and woodworking.

Classrooms vary in design, according to their building. Those located in 100-year-old converted homes are charming, with bay windows, French doors with glass and gumwood trim, and hardwood flooring. The elementary building is new, and the classrooms are more traditionally appointed. All have windows, bookcases, shelving units, cupboards, bulletin boards, a whiteboard, and at least one desktop computer, although laptops are always available, and all senior students take part in a one-to-one laptop program. Until the end of grade-three, students have a hook and shelf in the hallway outside of their homeroom class. By grade-four, full-length lockers are provided. Outerwear, knapsacks, and personal items are stored there during the day and discouraged from being brought into classrooms. Locks hang from the lockers but are often unlatched. This school does not have a culture of stealing. I'm told it is a rare occurrence. Lastly, students have an individual desk cubby or shelf in their homeroom class for storing school supplies.

The school year is divided into three terms. The first begins on the Thursday following Labour Day, in September, and continues to the two-week break in December. The second begins after New Year's Day and ends at the start of another two-week break in mid-March. The final term begins after the March break and continues to the end of the second week in June. Because this school is not semestered, all schedules remain the same from September to June. A school-wide daily schedule operates Monday to Friday, as shown in Table 3.1.

Each day begins in homeroom class, with quiet reading, followed by the national anthem. The day also ends in homeroom, with announcements and the recording of homework. Students remain at school for lunch, eating in their classrooms four of the five days. One day per week each division rotates eating together in the gymnasium, while their teachers meet and plan. There are two recess periods, one in the morning and one after lunch. Unless the weather is inclement, all students go outside to the courtyard playgrounds or to the nearby municipal park, for which the school holds a permit to use. Curricular subjects are scheduled by class in seven instructional periods during the day. They rotate on an eight-day cycle. Table 3.2 reproduces the day-two schedule for Terry's class.

Table 3.1 School-wide daily schedule

Time	Activity
8:30–8:45 am	Homeroom/DEAR (drop everything and read) National anthem
8:45–9:30	Period 1
9:30–10:15	Period 2
10:15–10:30	Recess
10:30–11:15	Period 3
11:15–12:00	Period 4
12:00–1:00 pm	Lunch/recess
1:00–1:15	DEAR
1:15–2:00	Period 5
2:00–2:45	Period 6
2:45–3:30	Period 7
3:30–3:35	Homeroom/homework

Table 3.2 Terry's class schedule for day two

Time	Activity
8:30–8:45 am	Homeroom/DEAR (drop everything and read) National anthem
8:45–9:30	Period 1: math
9:30–10:15	Period 2: music
10:15–10:30	Recess
10:30–11:15	Period 3: social studies
11:15–12:00	Period 4: French
12:00–1:00 pm	Lunch/recess
1:00–1:15	DEAR
1:15–2:00	Period 5: writer's workshop
2:00–2:45	Period 6: writer's workshop
2:45–3:30	Period 7: language arts
3:30–3:35	Homeroom/homework

With five days of school per week and an eight-day cycle, each week is unique, as the days of the week can be any scheduled day, from one to eight. The students' excitement is palpable when their favourite school day also falls on a Friday.

The child-adult ratio across the campus is low. Regular classes are capped at 16 students. There are two such classes in every grade. Grades two to eight also have a class capped at 10 students, for those requiring consistent academic support. These are not referred to as special education classes, although they are, but rather as *small classes*. Students in a small class have a modified academic program but are integrated with their peers for physical education, health, music, drama, and art. The low child-adult ratio is also a result of having specialty teachers, and support and resource staff available for counselling and remedial or enrichment work. Parents and teachers primarily, but occasionally students themselves, directly contact these professionals with regard to a range of specific needs.

These services are mainly supported by tuition and incidental fees. The total amount for families is approximately 23,000 dollars Canadian per student, per school year. As a result, and with very few exceptions, the students are from privileged families. They have access at home to books, computers, and private tutors and are supplied by parents with a wide range of learning materials. They travel with their families for pleasure and in many cases have more than one home. They are appropriately clothed in all seasons and for all weather, although most days uniforms are worn, either formal or gym. One can reasonably assume they come to school well-nourished and with ample food for lunch. Parents are involved and interested in the school, supporting its programs financially with donations and volunteering and participating in special activities throughout the year. They are educated and speak English fluently and, thus, are capable of helping their children with homework, communicating with teachers and staff, and advocating for their children when needed.

Such extensive supports and resources have enabled Middlevale to pursue a holistic approach to education that addresses the whole child: "While attending to each pupil's academic development, [Middlevale] is also concerned with the 'whole child' [quotations in original], striving to maintain a balance of a child's physical, emotional, and moral development" (school website, June 2010). The school's ability to individualize this approach and "to do what is best for each child" (school brochure, 2009/2010) is facilitated by its size. Small schools are more readily able to foster relationships between the adults and children and to personalize the learning environment (Strike, 2008). Accordingly, the staff joke that the Head of School, Sean Patrick, not only knows every child, every sibling, *and* every parent by name, but he also knows the family pets. While resources and school size are assets, they do not inevitably result in students' moral growth and development. This is more directly nurtured by particular programs, school structures, and activities.

CHARACTER DEVELOPMENT PROGRAM

Middlevale's Character Development Program gives teachers and staff a script and the incentive to infuse moral values into the environment and monitor their own and students' attitudes, behaviours, and choices. Fenstermacher, Osguthorpe, and Sanger (2009) allege:

> Having an explicit moral program or curriculum within the school signals the teachers that moral matters are an important part of the work of teaching, and attention to such matters will be valued by colleagues and superiors. The adoption of a program of moral content, whether in the form of a "pre-packaged" curriculum, or as an organic product of the mission and philosophy of the school, served as a form of permission-giving to the teachers, as if to say, "Moral development is valued here; we want you to attend to it" [quotations in original]. (p. 12)

Middlevale's program was organically developed. However, it was not a product of the school's mission or philosophy, as Fenstermacher, Osguthorpe, and Sanger (2009) contend, and as Salls (2007) recommends: "A school's commitment to education in virtue should be reflected first of all in the school philosophy statement, next in the school mission statement, and lastly in the character education strategy itself" (p. 96). Rather, it grew from a philosophical and practical shift in the school's approach to discipline, which Sean initiated in his role as Assistant Head of School, prior to becoming Head of School. In his words:

> I wanted to change our approach to what was essentially student discipline. I wanted to show kids how to make the right choices. This began with me understanding that I wasn't really doing a very good job of changing behaviours of kids by doing punitive things. "If you do that again you're going to have a detention. I'm going to call home. I'm going to suspend you".

Sean was reacting against the contingency-based classroom management and discipline strategies, derived from behavioural psychology and promoted throughout the 1970s and 1980s. Rewards, punishments, and incentives enabled short-term control of student behaviour but paid little regard to social and moral learning (Watson, 2008). Sean's beliefs coincided, instead, with developmental psychology, which assumes that children are naturally predisposed to being cooperative and pro-social, and if provided with opportunities to acquire social, emotional, and moral skills, in an environment of goodness, they will increasingly do what is right and good. Thus, the developmental approach to discipline involves policies, procedures, rules, and guidelines that enable and encourage desirable behaviour; help students to learn from mistakes; preserve students' sense of worth, competence, and dignity; provide opportunities for students to practice being kind, fair, and responsible; and allow teachers to model goodness (Watson, 2003, 2008). Punishments and punitive measures are understood as counterintuitive and counterproductive. They may even be unethical given their intention to harm wrongdoers (Watson, 2003, 2008).

Sean's philosophical shift led to a new articulation of discipline:

> Our focus is on the positive side of reinforcement, rewarding appropriate behaviour. Expectations are set out as guidelines that students "can" follow, rather than lists of "can't" [quotations in original]. These guidelines are meant to encourage students to think and act responsibly, rather than encouraging avoidance techniques. (parents' handbook, 2009/2010)

Expressed in terms of moral development, discipline became "teaching thoughtful regard for the welfare of others" (school website, June 2010). This coincides with Goodman's (2006) assertion, "If schools take seriously the moral development of students, their discipline policies should be a conduit for moral instruction" (p. 213). The school did not, however, have a formalized, systemic, or systematic approach to moral instruction. Sean began to explore options. His search coincided temporally with the Ontario Ministry of Education's growing interest in character education, as noted in the previous chapter. Although Middlevale is not a publicly funded school and, therefore, not subject to the Ministry's 2006 mandate for creating and implementing a character education program, Sean was, nonetheless, inspired by the political and social climate. He collaborated with Terry to investigate an approach to character development that would suit Middlevale's nature and needs. The history of this program is recounted in their words, which I have woven together from statements each made independently of one another.

> SEAN: It was initially conceived of as a curriculum piece. Writing curriculum over the summer is a classic school project. We're going to have a team of three come in, and they're going to look at our curriculum, our science curriculum or social studies, or even smaller, our research skills curriculum. Somebody's going to map that out. And they look at best practice, and they come up with a bunch of skills. And then they come up with lesson plans about how you're going to teach these skills across the grades. And really, I saw character education like this, as being curriculum writing.

As he had for other projects, Sean circulated an email requesting a teacher volunteer. Terry responded. Her task was to research character education, what it was and where it was taking place, and to write a customized version for Middlevale that would be distributed to all teachers and taught in conjunction with the health curriculum.

> TERRY: Sean approached me in the spring of 2004 and asked, "What do you think about doing a little bit of research on character

education?" So, for about a year I was looking at different journal articles to do with character education, and I went to workshops.

It soon became apparent to both Sean and Terry, however, that a curricular approach was not appropriate for Middlevale.

TERRY: If you go to other schools, they actually have a program. And we thought about that. We thought long and hard. Is it worth doing an action program where you have a scope and sequence, like any curriculum strand? But we didn't think that was right for our community.

SEAN: Terry came and said, "I don't think I want to do what you think you want me to do. I don't want to write a grade-four curriculum or a grade-one to eight curriculum around character because it's not going to work. It won't do anything. There's no point". I know now that that is entirely the wrong direction to go because it can't be a stand-alone. It can't be an overlay. We don't want a shrink-wrapped version delivered—here's your grade-four respect worksheets, role-play, whatever. It has to be woven somehow right into the school. We wanted to weave this into the culture of the school with words, actions, behaviours, and characteristics, for all of the humans that come here every day.

SEAN: And so we stopped that project. But Terry and I continued to read together. We visited lots of schools. And this became a filter I looked through when I'd visit other schools. And out of all this came the notion of an ethos of values and development, rather than traditional character education.

This shift in orientation was signalled by a change in language, from character *education* to character *development*.

SEAN: And then, Terry and I worked together, cochairing the committee. We had a couple of parents on that committee; we had a few kids; we had other teachers. I think there were about 10 of us. And we met over the course of the year.

SEAN: I had come to understand that we needed to define the values that we wanted to have the children experience. I don't want them to just learn, but to experience, day-to-day, so they could develop those values in themselves. And I was very conscious that I wanted that to be a shared process, all of us together, or that there should be community input. A process of asking together, what is really important here? What do we want the children to emerge from here with? And we had focus groups in the school. And we did a lot of brainstorming. We had faculty meetings set aside for it and parent focus groups and questionnaires that were sent home.

TERRY: And then we had the character education committee come up with a list of values. But we couldn't choose too many. So what was the right number and which ones? I looked at a study of different cultures and different values. There were similarities. Respect and responsibility are common throughout the world, and not only based on religion, culture, or where you come from. They're just human. Humans are a community-based species, and in order to get along we have similar values.

SEAN: We were able to distil out the five values—respect, responsibility, integrity, compassion, courage—from which our program emerged.

TERRY: By defining the values there was a little more focus to what everyone was doing because the teachers were already randomly giving some form of character education. This way, everybody had the same language and same values from junior kindergarten and up.

TERRY: Then I asked people to share their materials with me, anything that related to character. This included the peacemakers' stuff and articles and programs that other teachers were using. I even interviewed teachers. We wanted to see what others were already doing, not to ask them to change but to share and build upon that. Grade-one, for example, still has their *I Care Cat* stuff, but they also do *Project Love* where they collect school items for those in need. Everybody seems to have adopted something. The grade-eights clean up the environment somewhere. Some classes go to another school and buddy up with a grade-two class for drama. There are new activities that are added each year. We've had speakers from Free the Children, and food bank people. It's growing. Everybody has the capability to make it their own now. That was what we wanted.

SEAN: One of the things we also said was it wasn't like this was a barren wasteland of unethical behaviour before we stepped up. It was to try to name and intentionalize a lot of the stuff we already did as a culture, as a school community, so we could do it better. "Here's what you're already doing that falls into these things. And here are some other things that we've heard". Or, "Here are some things that your colleagues are doing. Here are some DVDs and books that we liked. Here are ways to do it that are really appropriate for a 6-year-old, a 4-year-old, a 14-year-old, and a 45-year-old". I include the adults because we can't talk about values with the children in full authenticity unless we're actually behaving accordingly ourselves.

SEAN: That all begins with the adults. I now know better what those [values] are because I've taught them. I've been working with the children, and I've been working with my colleagues. What does this look like in our school? What does this look like in our lives? We have been living it. And you stop and you rest on those words. If we have to have difficult conversations with parents, like the conversation that your child will have to go into our special education program, or this may not be the right school for your child, we say, we have to have the *courage* to do this. Our responsibility is to this child. A colleague died last year and we talked about compassion. I had a conversation just this morning with a colleague about very difficult things she's been through with her family. We talked about our school values of courage and integrity. We talked about how we don't just *talk* about courage and integrity, as the adults here, but we also have to *be* them. You rest on those words. But they're not just words because you build up against them.

SEAN: I think that this more clear articulation of values in our school community has allowed the adults, who really determine the culture of the school, to hold themselves differently and conduct themselves differently and be somehow more ethical, moral people in their practice with children. And it's going to rub off on the children. Any shift in school culture begins with the adults. We do far more for our children by what we do than by what we say. This is *the children are watching us* stuff.

SEAN: And we always come back to our values. We come back to the words. To the children we say, "You need to have the courage to not stand by while this is happening in your classroom or on the playground. You need to have the respect". Or, "I challenge you to have the respect. What would compassion look like in this situation? We are a place to talk about compassion. Now's your chance to exercise some. What am I going to see?" And the nice thing is, all of the children know what those things look like. They know. Integrity is a hard one. But every kid I know in the school has heard the words and knows what they mean in an age-appropriate way. And when the values do come around again and we are faced with opportunities, when it's decision time, then hopefully some of what we talked about resonates.

SEAN: So in the end, we did not roll this out as an add-on, a curriculum piece. It's not a curriculum piece. It's a cultural piece that permeates all aspects of our school life, for the children as well as the adults in our community. It's human to human to human.

In this historical account, Sean identifies the Character Development Program as both a means of defining the school culture, or way of life at the school, and a means of moral education for students. These two functions are related. Strike (2008) asserts that the moral norms of school culture have unintentional consequences related to providing a moral education for students, and that the key to a good moral education lies in building healthy school communities. Salls (2007) agrees: "The basic context of character education in schools then, is the environment of the school as a whole. . . . Everything that occurs within and around the school can be the occasion of character education" (p. 100).

Several years after its 2005 implementation, I am able to observe examples of how the Character Development Program has fulfilled Sean's vision to be relevant beyond the classroom. Administrators frequently reference the school values of respect, responsibility, integrity, compassion, and courage. At assemblies, for example, they say versions of the following: "It takes a lot of courage for people to stand up in front of all of you, and make a presentation. Even some teachers need courage to do this"; and "If you are sitting quietly and still, listening, clapping, and laughing when it is appropriate, you are showing respect for the person who is up here talking". Achievements in community outreach are often discussed in terms of responsibility and compassion, and achievements in sports are discussed in terms of courage and integrity. Further, when awarding silver and gold pins, Sean noted that recipients demonstrated personal integrity by honouring commitments made to a range of activities in leadership, athletics, and arts. Values messages and character expectations for students are similarly communicated to parents. The passages below are modified from two newsletters:

> At Middlevale we believe that the development of both MORAL and PERFORMANCE character are essential for children to succeed in life. . . . Character strengths such as empathy, fairness, trustworthiness, generosity, and compassion are aspects of our capacity to love. These qualities make up our MORAL character . . . character strengths such as effort, initiative, diligence, self-discipline, and perseverance constitute our capacity to work. These qualities make up what we call PERFORMANCE character. (2008)
>
> We challenge students to become better human beings, to learn from their mistakes, but also to accept individual differences. It's about learning to help and support one another and to look after each other. . . . We believe that children develop character by what they see, what they hear, and what they are repeatedly led to do. . . . The many adults we have working here every day with your children live and breathe these character values through their pedagogy as well as their daily interactions with your children. (2010)

The students have responded to this ubiquitous messaging, spontaneously and habitually displaying behaviours consistent with the school's values and the goals of the character program. On their own initiative, they hold doors open for others to pass through. They say, "You're welcome" when thanked and "Thank you" when the door is held for them. Students pick up items that have fallen, regardless of whether they were responsible or to whom the items belong. If a conflict arises on the playground, others rush to help and offer support or to get a peacemaker or adult. Students are respectfully engaged during assemblies, clapping, laughing, and listening. They are helpful to visitors and guests, in official roles as tour guides and buddies and unofficially if someone simply looks confused or lost. When such behaviours are not exhibited, or behaviours are disrespectful, unkind, and irresponsible, for example, all adults at the school are expected to intervene with a reminder or a more significant consequence. These generalizations do not apply to every child, every adult, or every incident. Yet, I did overhear an older student contextualize the school's ethos as follows: "When you are playing soccer at my other school and you make a bad shot, the kids say, 'You suck!' When you make a bad shot here they say, 'Nice try; good shot'".

While proud of the progress made, Sean, nonetheless, acknowledges that the teachers are at varying levels of awareness: "Some teachers get it more than others". Terry agrees:

> Not everyone's going to have moral education, character education first and foremost on their mind. They will teach the five values that we stand for as a school. But whether that becomes a daily conversation or goes through their whole curriculum, I'm not sure. My expectations are always there for that, in every subject they do. But people may not do the same thing. Or people may not necessarily do it, until it's time for health class. I carry on and do what I can in here and out there, even in the hallway. It's hard to know what somebody else really does, though, in their own class.

There are still incidents of teachers yelling at students, sometimes reducing them to tears. There are also stories of teachers not taking the time to fully understand a situation and implementing what some students feel are unfair consequences, including group punishment. On occasion, a teacher will bring a laptop to an assembly and work, rather than engage in the proceedings. These are not in accordance with the school's values of respect, responsibility, and compassion, in particular. "It's a process and we're working on it", Sean admits. He continues:

> I think we have 100% buy-in from the teachers that this is the way to be in this community and that we're not just about the delivery of curriculum here. This is why these people choose to become teachers. But

some of them are, like everything else, a little further along the learning curve and a little more open and available to colleagues who are putting in place some of the coaching skills to support character development. We all need some coaching, to be coaches.

Sean often evokes the metaphor of coach to describe the teachers' role in the character development of students. Fittingly, he provides teachers with the opportunity to participate in a coaching workshop so they may acquire and refine related skills in listening deeply, asking impactful questions, providing meaningful feedback, cocreating relationships of trust with students, communicating effectively, and holding the students' agenda, rather than their own. Although not mandatory, the teachers are strongly encouraged to participate. Terry could not think of a single teacher who had not done so:

> Pretty much all of the teachers have gone through the coaching workshop. We go through several days of it in the summer, and we learn the steps to take. We practice with each other first. It's one of those self-awareness things. It's great for educators and parents alike. And certainly with adult-to-adult relationships it's good, as well.

Terry notes specific benefits to her own practice:

> The coaching workshop was very eye-opening in the sense that this is what we normally do when we listen. When we listen to adults we interrupt a lot, and we put in our judgements to situations. If someone was telling me a problem, I did judge. Also, before the coaching workshop I used to be in such a hurry when I listened. If they had a problem after recess I always tried to hurry things up, and before they finished telling me their story I used to interrupt and say, "Is this what happened next?" because I was trying to hurry them along. And after the workshop I learned that it has nothing to do with what I think about the situation. I learned to step back and take myself out of it and just listen. . . . I learned to be a better listener, and I learned to ask questions that are pertinent to what the person's telling me and not to just make assumptions about the problem. So now I always say to the kids when I'm listening to them, "I'm not making a judgement here. I'm just asking".

The workshop also helps to maintain a school-wide continuity, with respect to how teachers guide students, year after year. Terry explains:

> The strategies for solving their social problems, they've already seen because they had problems last year. What happened exactly, I don't know. But I'm pretty sure they were trained to talk about it. They probably had time to talk about it one at a time, too, and were told not to interrupt and just to listen to everybody's versions, then come up with something

together. . . . I see them for this year, but all the time that they have been at the school, somebody's done one thing or another to support what I also do. Then they just go through me and move on to somebody else.

From Sean's original idea to create a piece of curricula supporting a developmental approach to discipline, therefore, evolved a comprehensive character development program, which became a central concept in the school's philosophy and mission; a defining feature of the school's culture; and a focus for both teacher and student development. This is expressed in the parents' handbook (2009/2010), as follows:

> The character program was created to help ensure that shared values of good character are infused into the curricular and co-curricular programs of the school, and are visible in the actions and daily interactions of all members of the school community. It intentionally cultivates the development of Respect, Responsibility, Integrity, Compassion, and Courage in the members of our community, so they foster a conscious, lifelong commitment to themselves, other people, and the Earth. These five qualities are shared values of our community. They transcend individual differences and act as a touchstone to many other important ethical and moral attributes of good character. The development of character is a shared responsibility, in which the school takes a supporting position to the parents and families of our students.

Peacemakers and Community Outreach Programs

The Peacemakers and Community Outreach Programs are expressions and applications of the Character Development Program, promoting its values and providing opportunities for students to practice associated behaviours and acquire desirable assumptions and attitudes. Terry helped to develop both programs and continues to coordinate them at the school level. Peacemakers is a two-year leadership commitment, open to students in grades four and five. During recess, peacemakers patrol the playground in pairs, offering peer-mediated conflict resolution support, as needed. Approximately 40 students participate. Most are girls. In the first term, the grade-five students patrolled the playground, while the grade-four students trained. Terry and Wendy Bell, the co-coordinator, conducted training sessions every Monday at lunch. Students watched related videos, interviewed former peacemakers, and practiced the peacemakers' protocol with role-playing activities. In January, the grade-four students joined the grade-fives on patrol.

The peacemakers' protocol is guided by a script, which peacemakers carry on a clipboard. Developed for the Middlevale community, it is consistent with the values and assumptions of developmental discipline and the teachers' coaching workshop. Related to developmental discipline, peacemakers are expected to be supportive and caring; to maintain a democratic process; and to preserve the autonomy and competence of the students they are helping (Watson, 2003, 2008). Related to the coaching workshop, peacemakers are expected to listen carefully, ask clarifying questions, and give everyone involved a chance to express themselves and be heard. Accordingly, the protocol begins with peacemakers offering their service: "We're peacemakers. Would you like us to help you solve your problem, or would you like to work it out with a teacher instead?" If the children decide to continue with the peacemakers, as they often do, each child is given an opportunity to tell their version of what happened, without interruptions. When everyone has finished, peacemakers clarify the details and begin problem-solving: "Let's see if we are ready to problem solve now. What suggestions do we have?" Together they work through different solutions until everyone is satisfied. Of obvious benefit to those who are assisted, this process also has a morally educative benefit for peacemakers. Howard (2005) claims that one's moral judgement and reasoning improve by helping others to work through problems and find solutions. Further, Johnson and Johnson (2008) note that peer mediators necessarily learn to express moral values of fairness, compassion, empathy, and honesty.

The Community Outreach Program is Middlevale's version of service learning, defined in the literature as an experiential pedagogy combining classroom instruction and community service activities, such that students apply what they learn in class to real-world situations and problems; perform acts of service to the benefit of others; and reflect individually and collectively on the experience of doing so (Billig, 2000, 2009; Furco & Root, 2010; Hart, Matsuba, & Atkins, 2008). Advocates claim two morally positive outcomes—meeting global or local community needs, as well as furthering students' moral learning, growth, and development (Hart, Matsuba, & Atkins, 2008; Howard, 2005; Howard, Berkowitz, & Schaeffer, 2004; Lickona, 1991; Ryan & Bohlin, 1999; Schuitema, ten Dam & Veugelers, 2008). Grounded in the school's values, Middlevale similarly situates service as *social* action and *soulful* action:

> Social action involves tackling important issues in our community, our country, and around the world. Soulful action involves activities that focus on personal reflection and self-transformation. Emphasis is placed on raising students' awareness of the world around them and challenging them to take the initiative to make a difference. While fundraising is an important part of assisting worthwhile organizations, cultivating a sense of volunteerism and social engagement is the most important outcome of the Community Outreach Program. We believe that students' reaching their potential includes becoming more engaged and responsible citizens. Community outreach allows students to develop skills, and provides the tools for meaningful and active community service. (parents' handbook, 2009/2010)

Accordingly, all students are involved in several service activities throughout the school year. Many activities are school-wide, including sending postcards to war veterans on Remembrance Day, donating books to the Bookshare Program, raising money for a variety of charities, and filling hampers with clothing, toys, and food items for families in need. Other activities are by grade or division. In the early grades, for example, students created and sold buttons for the World Wildlife Fund and wrote letters or drew pictures for residents of a seniors' facility. These activities are more characteristic of community service, as they do not usually derive from or initiate academic study. In the higher grades, however, activities such as planting trees, testing the quality of water, and participating in nature cleanups tend to support the science and social studies curricula, thus, qualifying as service learning. Further, and also characteristic of service learning, older students are expected to assume leadership responsibilities in determining, organizing, and implementing service initiatives (Morgan & Streb, 2001). The school's intention to scaffold the program's learning experience is explained, as follows:

> As students progress through the school, they are expected to take on more responsibilities for organizing their community outreach activities, and determining where their volunteerism will be directed. From junior kindergarten to grade 2, community outreach activities are facilitated by the homeroom teacher and take place in the school. They emphasize class discussion, cooperative decision-making, and raising awareness. From grades 3–6, students will decide on a class activity that will allow students to engage personally with the local or global community. Grades 7 & 8 students will aim to participate in structured community outreach opportunities. (parents' handbook, 2009/2010)

As with the school values and character program, community outreach is relevant to the entire school community, including teachers, staff, and parents. Teachers and staff participate with the students and role model desirable attitudes and behaviours. Parents are informed of initiatives through newsletters and are expected to support their children's participation and to participate themselves, as the following example illustrates:

> This Holiday Season, we invite you to continue the tradition of giving. The school will be making donations to three local organizations, which serve those in need at this particular time. In addition, the school will be once again sponsoring 24 local families. Within their class, or as a class pair, students will be given biographical information about a family in our neighbourhood who they will "adopt." Each family will be provided with food, grocery vouchers, warm clothes, and small gifts to brighten the children's holiday. Cards and letters to the family are also included in the hamper. (newsletter, 2009)

Parents are also encouraged to honour teachers and staff by making donations in their name, rather than purchasing personal gifts.

> While gifts to teachers are always appreciated, a donation to one of these causes in honour of a member of the faculty is also valued. Our faculty are supportive of these causes and appreciate this thoughtful gesture. (newsletter, 2009)

The Character Development, Peacemakers, and Community Outreach Programs, although continuing to evolve, are firmly entrenched in Middlevale's structures and culture. Subsequently, the Character Development Program, in particular, has eclipsed the influence of its creators. Terry and Sean no longer assume direct responsibility but continue to be a resource for Jonathan Blakely. As Assistant Head of School, Jonathan is also responsible for campus discipline. Hence, the conceptual and practical connections between character development and developmental discipline are sustained.

SCHOOL STRUCTURES AND ACTIVITIES

Middlevale's ability to nurture the holistic growth and development of each student is also enabled by a cohesive, inclusive, and caring school culture, which is shaped by particular structures and activities that bring the school community together for shared experiences and organize students, teachers, and staff into a variety of groupings. Characterized by relaxed informality, positive energy, and a spirit of celebration and camaraderie, school-wide assemblies represent the most frequent and consistent shared experience. Every Monday morning at 8:45, students, teachers, and administrators gather in the gymnasium. The national anthem is sung together, led by one of three dedicated music teachers. Student leaders assume responsibility for the agenda, calling on students first, then teachers and staff to make announcements and presentations, to show videos or slide shows, and to celebrate collective achievements, particularly those related to community outreach and athletics. Teachers bring their morning coffees and sit together on benches along the periphery of the room. Students sit on the floor with classmates and other peers and are minimally supervised. Rarely is disciplinary action necessary, and if it is, a reminder, small adjustment in seating position, or removal of a distracting object usually suffices. The students are trusted, respected, and empowered, and they respond by being steadfast, respectful, and self-regulating.

Special activity days (SAD) occur three times during the year and similarly bring the school community together. Regular class schedules are suspended for an alternative, cocurricular program, organized around a particular theme. In my year of fieldwork, the themes included Remembrance Day, cultural diversity, and environmental sustainability. Each was launched in a morning assembly, distinct from the Monday assemblies. For Remembrance Day, the gym was arranged in rows of chairs, lengthwise in front of a stage. Teachers sat with their students. The assembly was longer and more sombre and included external speakers and guests. In contrast, opening assemblies for the other two SADs resembled pep rallies. On cultural diversity day, for example, two representatives from each grade marched into the gymnasium carrying homemade flags from around the world. The song known as *Wavin' Flag* (Universal Music Group, 2010) played loudly in the background. Students immediately rose to their feet and began dancing, clapping, and singing along. Teachers and staff joined them. The remainder of the day, for all three SADs, involved rotating through a variety of educational, explorative, and interactive experiences, often led by guest speakers, presenters, and performers. For example, Remembrance Day included a session on marching, where students learned commands from a former soldier. Cultural diversity day included dancing and singing with a steel drum band from South America and storytelling in the French Canadian tradition.

For these three days, and during other activities throughout the year, members of the school community participate with their *house*, and not according to divisions, grades, or classes. Upon joining the school, students, teachers, and staff are assigned to one of four houses, named after an inspiring individual and moral role model who demonstrated personal determination and courage and was responsible for significant, positive, and lasting contributions to society (school website, June 2010). Each house is also associated with two of the schools' values: respect plus either responsibility, integrity, compassion, or courage. These connections strengthen the Character Development Program and provide many opportunities to reinforce moral values. Further, in assigning houses, there is an effort to maintain diversity regarding the age and gender of students; the experience, subject speciality, and gender of teachers; and the roles and responsibilities of staff members. This facilitates otherwise unlikely cross-campus contacts and relationships and places several adults and older children in a position to model pro-social and moral conduct and behaviours. One such opportunity occurs during SADs, as older students are expected to lead and care for younger students during rotations.

Finally, cocurricular and extracurricular teams and clubs bring together students teachers and staff, according to shared interests, talents, and experiences. Students may explore opportunities in leadership, athletics, arts, and technology. Leadership opportunities include primary helpers, game referees, student ambassadors, learning buddies, peer tutors, tour guides, peacemakers, and student council members, as well as captains for assemblies, athletics, community service, sustainability, literacy, spirit, technology, communications, and yearbook. Students in grades four through seven are encouraged to assume a leadership role as a means of practicing accountability and responsibility for oneself and for others. Leadership roles are assigned for every grade-eight student. Competitive and non-competitive athletic opportunities are varied.

Soccer, basketball, ice and field hockey, cross-country running, ultimate Frisbee, volleyball, and track and field teams compete in tournaments with other independent schools. Tennis, T-ball, badminton, walking, intramural sports, flag football, curling, shinny hockey, and pleasure skating clubs are non-competitive. Arts clubs include arts and crafts, concert and jazz bands, song writing, theatre, props and set design, ballet, tap and jazz dance, choir, writer's craft, needlepoint, and sewing. Lastly, technology clubs include computers, photography, and iMovie.

Given so many options, all students are encouraged to participate in several activities. Some operate for the entire school year, while others are only offered in winter or a particular term. A school-wide assembly is held at the start of each term to introduce new offerings. With excitement and controlled chaos, the students move around the gymnasium to different booths, inquiring about activities and signing up for as many as they like. The following week, homeroom teachers determine with each student what is logistically possible. Most teachers and staff also participate as leaders and helpers. A technology staff member, for example, co-leads the computer club; one of the custodians helps with props and set design; administrators play shinny hockey; the librarian leads writer's craft; and Terry co-coaches the field hockey team. The campus is literally bustling with activity and energy from 7:30 in the morning until 5:30 in the evening, Monday to Thursday, and 8:25 until 3:35 on Fridays.

A PERFECT "STORM"

Ryan (1993) claims that schools positively impact the character of students when the following is established: a widely known mission statement; a comprehensive service program; school spirit and healthy intergroup competition; an external charity or cause that is supported; an award system to recognize effort, achievement, and contribution; expectations for students to be moral role models and exemplars; displays of heroes; and rituals for community celebration. Given this, Middlevale is undoubtedly an ideal school for teachers who prioritize the moral education of students, the positive equivalent of a perfect storm. It is fair to assume, therefore, that Terry's practices, as detailed in the next two chapters, are liberated and enabled in this environment. Biesta and Tedder (2006) assert, "The achievement of agency will always result in the interplay of individual efforts, available resources and contextual and structural factors as they come together in particular and, in a sense, always unique situations" (p. 137). This might cause educators to question the relevance, generalizability, and transferability of this portrait, particularly if their own schools are not similarly focused on character development or the expression of pro-social and moral values.

Yet, it should be remembered that Terry was a leader and active participant in creating this environment, generating commitment from the school community, motivating staff and teachers, acquiring and distributing resources and knowledge, facilitating a process to identify school values, and helping to create the Character Development, Peacemakers, and Community Outreach Programs. One might also assume, therefore, that Middlevale's environment was liberated and enabled by Terry. This is an interesting claim that infers a broader role for teachers, beyond their own students and classrooms, to be influential school-wide. In addition, there are approximately 30 faculty at this school, who are subject to the same teaching and learning environment as Terry. While their teaching and professional reputations are generally excellent, Sean identifies Terry as extraordinary among them, particularly in regard to moral education:

> Terry was teaching grade-five, right next door to my office. So I knew she was a brilliant teacher, a great teacher. Her kids learned. No matter where we put her, the children she taught learned and were engaged. She has good teaching techniques. She plans well; she connects well with the children; she has great activities for them to do; she assesses well; she writes good report cards. She definitely had the knowledge base. But what I also know about Terry is that being an ethical practitioner is incredibly important to her. It isn't just a belief. It's something she does because it is the right thing to do. It is the marriage of her own personal ethics and what she will accept of herself and sees as her responsibility, with this deep love for the kids. There are children here, and children no longer here because they've graduated, that when I look at their journey through the school, their time with Ms. Kennedy stands out. She was the one who would not let it be okay until they were that best version of themselves. She spotted something in them and said, "You can do better, and I'm here to help you".
>
> Terry has really transformed lives. Not tons. But those individual lives that maybe could have gone one way or the other. She relentlessly held those students honest to the values. Bullies bully usually because they don't feel very good about some aspect of themselves. I've seen kids like this, and I wasn't always positive they were going to be able to stay at this school. She'd say to them, "We're going to talk about this at the end of every recess. We're going to talk about this, and then I'm going to talk with your parents openly and honestly about what my worries are here, where the opportunities are, what I intend to do about it. I'm not going to let this go. You have better in you than this, and I'm going to be the one to hold you to that". It's easy to say, "That's okay. He's going to get it eventually". But she'd make sure that he got it that year. I could list the four or five kids that I know she was the single most important influence in their ethical development. Grade-five was so different than grade-three for them, and it's because of what happened in grade-four when they had Terry. And sometimes she doesn't make the big difference right away. But three years later all that drip, drip,

drip pays off. A better version of that child emerges out of their time with Terry.

Reflecting on this, Terry admits:

> I think I may sweat the small stuff more than other [teachers] and more regularly than others. I think maybe others would not have dealt with all the moral issues on a regular basis, as I have. I try to address as much as possible, all the little issues. Not every issue, but all the ones they're ready to address and that I'm ready to address. I think some people probably would have ignored them and just moved on.

Lastly, the specific practices Terry employs are consistent with reports from other empirical work (Campbell, 2003, 2004; Campbell & Thiessen, 2000, 2001; Fenstermacher, 2001; Jackson, Boostrom, & Hansen, 1993; Richardson & Fenstermacher, 2000, 2001). These studies were conducted in a variety of school and classroom contexts, demonstrating that the practices themselves are robust and potentially transferable.

It cannot be concluded, therefore, that the school environment is entirely responsible for the moral education Terry is able to impart to students, and the favourable environment does not make this portrait of Terry irrelevant to teachers in other school contexts. This is, undoubtedly, an exemplary case study. Yet, the classroom practices are real and possible for teachers who commit to the moral education of students and consciously and thoughtfully apply their personal sense of morality to their teaching role and responsibilities. Thus, transferability is in the hands of readers who will determine for themselves what is contextually reasonable, individually comfortable, and professionally desirable.

TERRY'S CLASS

Terry's classroom is located on the second floor of the elementary building. Large windows create a spacious feeling, although the room is of modest size. A clothesline diagonally crosses the ceiling, from Terry's desk at the southeast corner, to my station at the northwest. Student work regularly dangles from coloured pegs, back-to-back for double capacity. Two bulletin boards along the west wall are also covered with student work, usually from language arts, science, and social studies units. Terry reserves the bulletin board by her own desk to post professional items, such as schedules, lists, and memos, and personal items, such as photos of friends and family. Progressively during the year, it also fills with notes and hand-drawn pictures from students. The bulletin board by the door to the hallway holds a world map. A second door leads directly to the other grade-four classroom. The two teachers and their students readily flow back and forth, sharing resources and collaborating on activities. From 11:00 in the morning, when the sun begins to stream through the south-facing windows, a rotating prism fills the room with multiple floating rainbows. Adding to this colourful and lively decor, Terry often displays a vase of fresh flowers on the windowsill.

Student desks are not affixed to the floor, and different arrangements can be achieved. There were two layouts during my time there: one with four groups of four desks, and the other a U-shape opening toward the whiteboard. These are diagrammed in Figure 3.1.

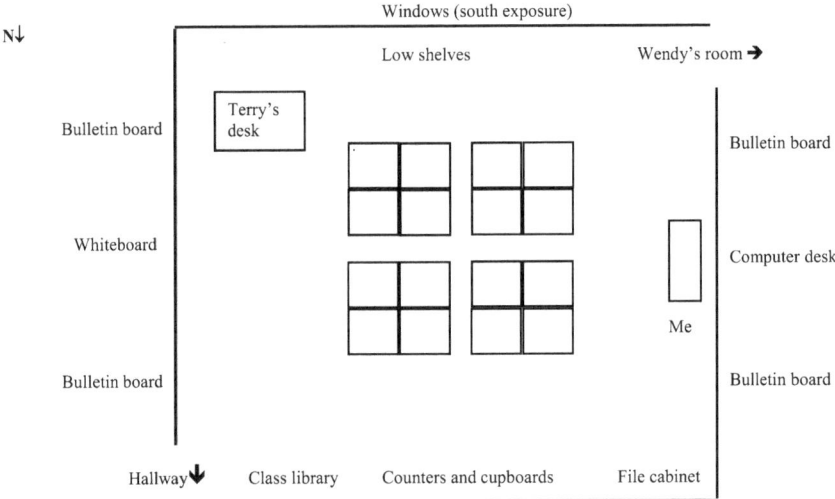

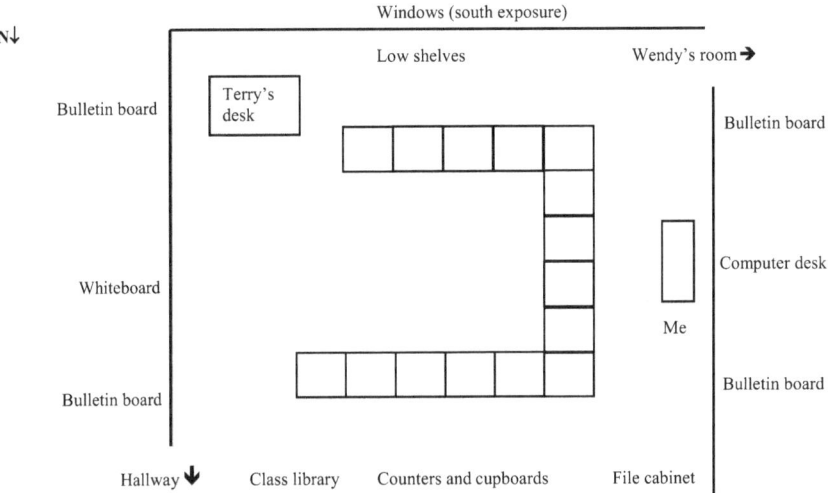

Figure 3.1   Desk orientations

A magnetic whiteboard fills most of the room's east wall. It is the hub of classroom activity, bearing key information, reminders, notes, and the daily agenda, in blue, black, green, and red erasable marker. This is also where lessons are taught. Although the details are changed regularly, their location, as shown in Figure 3.2, is intentionally consistent, enabling students to easily orientate and become self-directed. In support of this, Terry often replies to housekeeping inquiries with versions of "You know where to look for that".

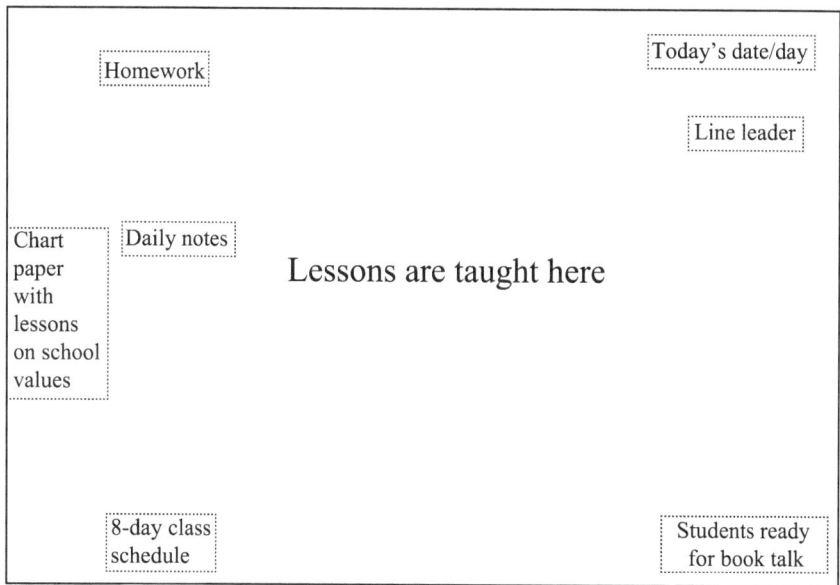

Figure 3.2   The whiteboard

Several posters adorn the remaining wall space. Two are explicitly moral: *Kindness is showing that you care*, and *Respect others. Respect yourself.* Other posters are values-based and implicitly moral. For example, *How you play shows some of your character. How you win or lose shows all of it* is about good sportsmanship and conduct and, thus, implies respect and fairness, among other moral values. *The expert in anything was once a beginner* implies perseverance, grit, and humility; *You'll always miss 100% of the shots you don't take* implies risk-taking and courage; and *Whatever you do, work at it with all your heart* implies diligence and responsibility. Terry also posted two inspirational magnets. One contains the following message by Marcel Proust: "Let us be grateful to people who make us happy; they are the charming gardeners who make our souls blossom". The second contains a message by John Fischer:

> The essence of our effort to see that every child has a chance must be to assure each an equal opportunity, not to become equal, but to become

different—to realize whatever unique potential of body, mind and spirit he or she possesses.

Terry did not discuss these postings with the class as a whole but readily identified them to individual students as the messages became relevant to their personal situations. She explains:

> Some years I remember to do a little walkabout and point out to the students all the posters and messages. But most of the time they're just there, if I need to refer to them. So for example, if the students are giving up and they're getting frustrated or something is too hard, there are two posters—*The expert in anything was once a beginner*, and *You'll always miss 100% of the shots you don't take*. Sometimes I've had classes where maybe one or two don't sign up for any clubs at all. And I really want them to sign up for a club. So then I say, "You'll always miss 100% of the shots you don't take. If you don't take that risk you'll never know whether or not it's something that you're good at, something that you'll enjoy, something that will give you more friendships. You never know". So there are moments when I refer to them.

Mostly, however, Terry allows the students to draw their own meanings and find their own inspiration. On the first day of school, one of the girls did just that. Independently copying each phrase into her doodle book, Frances explained how these messages made her feel good. Terry relates to Frances's desire for keeping personally relevant messages close at hand. A worn photocopy of a poem by Louise Cullen, entitled "ABC's for Our Children", is loosely tacked to the bulletin board directly above her desk. It expresses 26 creatively crafted guidelines for interacting with children, one for each letter of the English alphabet. Terry identifies the poem's relevance to her practice:

> Although I don't look at my bulletin board regularly throughout the day, I always see something from it during the week that reminds me of the important things. That's why I keep ABC's up on my bulletin board. Sometimes also, parents need parenting help. And I thought that maybe that's something I can offer, if I think that somebody needs it. I just think about one line at a time: *dream about what they want to become; question things so they'll learn to wonder*. One thing per day. It's lovely. I just think, have I done that today? It's a reminder for me. Have I done any of that today? This week?

These postings, exhibiting both moral and non-moral values, connect the classroom's physical space to its culture by signalling Terry's expectations for a classroom life characterized by kindness, care, respect, sportsmanship, fair conduct, risk-taking, courage, grit, humility, perseverance, diligence, responsibility, gratitude, happiness, equality, and equity, among other values. This is intentional:

> I think the [physical] environment helps them. I don't think they would put it that way. I don't think they know what it is when they come in here and why it works the way it does—not that it works perfectly all the time. And if you were to ask them they wouldn't be able to verbalize what it was. But the atmosphere is conducive to being a certain way. This is the way that I run it. This is the way that I am. When you come in here, this is what you need to learn about being in this classroom.

Signalled by these values, Terry's students learn to express their individual sense of self and a sense of community. Risk-taking, grit, humility, perseverance, and diligence contribute to a sense of self. Kindness, care, sportsmanship, respect, fair conduct, gratitude, equality, and equity contribute to a class community. Happiness, responsibility, and courage contribute to both. This observation is developed in Chapters Four and Five.

The resulting classroom culture is active and noisy by traditional standards. Students are encouraged to be themselves, take initiative, and interact with each other on a variety of tasks and responsibilities. Except during presentations, lessons, and tests, they come and go at their own discretion, move from desk to desk to collaborate and help others, access resources throughout the classroom, sharpen pencils, drink from water bottles, visit the washroom, or use the computer. If these behaviours become disruptive and compromise the ability of others to accomplish tasks at hand, however, Terry intervenes with values messages related to respect and fairness or by temporarily curtailing the freedoms, saying, "You are all too noisy right now. I need you to return to your seats and quiet down for a while. Then we can try again". Class discussions are similarly energetic and lively, with students encouraged to freely participate. Terry also monitors these, as she explains:

> In the moment when things are moving and the thoughts are coming out, I don't mind [the students] calling out so much. It feels like the free flow of ideas, and I don't want to disrupt that. I tend to listen and to take it in. It feels like I need to move with the energy of whatever it is that's happening; I just let it go. . . . Now, mind you, in a free flow of ideas and a discussion they can also be really loud, at which point there's no point in discussing because we can't hear each other. I would have to put my foot down. I do say, "I need one at a time because I can't hear your ideas". Then there are times when there are too many people speaking. I say, "Did you interrupt? Was someone else talking?" So it just depends on whether or not I can hear and whether or not I can tolerate it.

Terry's tolerance for noise and chaos is higher than Ms. Laurie's, the school librarian, whose office is in the library, directly across the hallway. Ms. Laurie is nearly a generation older than Terry and of a traditional European culture where children are expected to listen more than talk, and adults are expected to control the behaviours of the children in their care. Terry appreciates this different viewpoint and sensitively accommodates her when possible:

> She has certain expectations of children's behaviour. Sometimes she thinks noise is uncalled for. For example, the other day they were coming in from outdoor gym, and they were all so noisy. But it was happy noise. She came out of the library and stood there watching them. And then she said, "That can't be Connor! All that noise? Can it all be from Connor?" That's her thing. But if it's happy noise I don't come down on them quite as hard. It also happened for the happy birthday song. I thought Connor was okay with making the noise that he did. So when she came in to discipline, I said, "Well it's just because we are excited for the party". I try to give a little bit of a reason, if it's called for, so that she knows that maybe there was a reason for it. So that she doesn't think that they are all bad. On the other hand, she has to live across from us during the school year. And if the noise disturbs her then she has a right to complain about it.

Despite occasional conflicts, Terry does not compromise her desire for a relaxed classroom atmosphere, using humour to ensure one is sustained:

> I don't want everything to be so serious and they don't smile. Funny things are going to happen. It's just part of the day. And sometimes humour serves to sort of lighten up the atmosphere when things are getting too heavy. I gauge it by how I'm sort of feeling myself. If it feels tight to me, I think that [the students], subconsciously, also are feeling kind of tight. They may not, but they may. So laughing about something makes it feel that little bit lighter.

Terry began the school year with 14 students. The other grade-four class, taught by Wendy Bell, was also below capacity with 13 students. There were only four boys in each class. In January, a new boy joined Wendy's class. In February, twin boys joined the grade, one in each class. Both classes maintained a total of 15 students for the remainder of the school year, with five boys and 10 girls in Terry's class, and six boys and nine girls in Wendy's. The *small* grade-four class, taught by a special education teacher, was all male, with seven students, and, coincidentally, another set of twins. Although each class was undersized, and unbalanced regarding gender, the peer group for this grade comprised 37 students, with 18 boys and 19 girls. The entire grade was regularly brought together for some subjects, field trips, special activities, and events. "The first thing that we wanted to do is make sure that they had a chance to work with the other class, for developing different relationships and a wider variety of people to work with", Terry explains. For physical education, health, and the occasional science and technology classes, boys and girls were segregated to enable a gender-specific approach. Terry supports this, noting that "gender becomes a more pronounced issue at this age than in earlier ages, both academically and socially".

Fairly homogeneous regarding socioeconomic status, Terry's class was diverse in other ways, with students of East Indian, Australian, Maori, Asian, Italian, South African, and German heritages. None is a first-generation immigrant to Canada, although the boy from Australia was visiting the country and would return home after three years. Regarding religion, most students are of Christian denominations, either practicing or non-practicing. Two students are Jewish, one student is half Jewish, and another student is Hindu. Family structures are mostly traditional, with a mother, father, siblings, and pets. One student's parents are divorced and remarried. Another student has two mothers in a same-sex marriage, and no father. Terry rarely mentioned these characteristics in class. While they might have been utilized to convey messages of respect, inclusion, and sensitivity, Terry considers them to be private and, thus, it is the students' prerogative to raise them, not hers. Diversity regarding students' learning styles and general abilities, however, is openly acknowledged and embraced for its direct relevance to the core purpose of schooling and Terry's pedagogy. For example, Alexander received remedial support in language arts; Noah was exhibiting difficulty with math and would be recommended for the remedial program the following year; Bonnie has attention difficulties; Pia has difficulty beginning tasks; Mary is an excellent writer; Kathy is exceptionally creative and dramatic; and Connor is gifted in math. It is these differences on which Terry builds a class community of acceptance, inclusiveness, respect, sensitivity, and helpfulness, among other moral values. Accordingly, she instructs the students to "get to know each other, what you each can do. And help each other out".

These students embraced me as part of their community. They chatted with me before and after class, and at lunch and recess. They sought my help with seatwork, science experiments, presentations, personal conflicts, bruised egos, and scraped knees. They wanted to know where I was when I was not at school and if I was going on the next field trip with them. They showed me items they had brought from home and included me in birthday and holiday celebrations. Although they remembered why I was in their class, occasionally asking about my research, they never treated me as an outsider. At the end of the school year, several of us said good-bye with hugs and tears. While this profile focuses on Terry, not her students, there is a customized aspect to Terry's practices, an individualized approach for each child. Thus, knowing the students as unique individuals is essential to the moral education Terry is able to impart, and fundamental to our understanding of her practices. For this reason, throughout the book I refer to the students with consistent pseudonyms.

TERRY'S BACKGROUND

Having situated Terry within her professional teaching context, I explore Terry's background by way of identifying the ideology that shapes her moral education agenda. Terry recounts her story:

> I was born in Indonesia. My family and I left when I was 8-years-old and moved to Libya, in North Africa. I went to an American International school there for five years. Then my parents sent my sister and me to boarding school in Texas. I stayed for three years, and my sister stayed longer to graduate from high school. I decided to go to high school here in Canada, after they finally allowed landed immigrancy. Then I completed a bachelor of arts from the University of Toronto in English literature and Japanese studies.
>
> I went to Japan to teach English and was there for two-and-a-half years. I didn't have any sort of inkling as to whether or not I was going to be a teacher when I went to Japan. Nobody in my family is a teacher. I wanted to go there to do the cultural study thing, to see what it's like. And I realized that I can do this. I can do this well, and I enjoyed doing it. I enjoyed the rapport you can have between student and teacher. And my students were adults. And teaching them English was just one-on-one at first. But I was enjoying the conversation and helping them to learn the language. Then later on in Japan, I taught junior high–level students, just tutoring small classes in English. Again, I was attracted to the connection that you can have with the students to help them learn. I loved waiting for them to go, "Oh, I can use it this way? Oh, okay. I get that".
>
> So that's when I decided to do teaching. I actually came into teaching late. I was thinking of going into business at first. I quickly switched out of that. I realized that I am more interested in a human career. And to help children is one way that I can have that. It just makes me more satisfied and fulfilled in my life, in being able to do something good for other people every day. I still like knowing that maybe I'm making a little difference for the next couple of generations that are coming up. And I find it really rewarding when I actually do get through to a child and there's that "ah-ha" moment. Or I see them developing later on, maybe two years after they've left me. And I know that they're still interested in some of the things we talked about together.
>
> I guess this desire to help partly came from my dad. He was a strict Methodist, and he made us go to Sunday school and church every week. He's a good, religious person. But he wasn't overwhelming in that way. It was still quite democratic in our house. We went to church. We went to Sunday school. Boarding school in Texas was an Episcopal boarding school, and we had to go to chapel every day. But at some point he

knew better than to actually be domineering about the whole thing. I did understand that I had a responsibility to be a good person, with a conscience to make sure that I contribute in some way.

So I came back to Canada and completed a Bachelor of Education degree at Queen's University. Then I went immediately back to Japan for another two years. When I came back for good, I spent one year as a teaching assistant at a private boys' school, for grades one to four. Then I went to another private school to start a junior kindergarten program. I was there for one year as well and then came back to the first school to teach grade-two. But it was a hard school to really feel comfortable in. It taught me a lot, but I was staying there until 6, 7, 8, sometimes 9 pm every day. It wasn't very good for having another life. I decided that's enough, and came to Middlevale. This is a much more relaxed school, with a friendlier and warmer atmosphere. People leave by 5 pm, and they have a family life. Or they are able to do other things.

I started in grade-three, covering a maternity leave. I was hired in November of 2002 and started with the class in January, right after their Christmas break. So from January to June was the grade-three mat leave. I continued in grade-three the next September until November when that teacher came back. I then took over a grade-five mat leave from November until June. So that was 2004. And then finally in the fall of 2004 I was fully hired to teach grade-four, and I've been teaching grade-four ever since. So that means that I've been teaching here since 2002. Seven years this fall. That's long. And while I was teaching here, I did a master's degree [in Education].

As Terry outlines these events, her philosophy for teaching and learning, her vision of classroom life, and her disposition toward moral agency begin to emerge. Terry's philosophy is revealed in her desire to stimulate "ah-ha" moments among students, with curricula and other lessons. In doing so, she prioritizes learning over teaching and places herself in the role of learning facilitator. Her vision of classroom life is echoed in the description of the school's atmosphere, as friendly, warm, and supportive of teachers' holistic lives. Terry's attraction to the human element of teaching, the personal con- nection teachers have with students, and the opportunities for teachers to do what is right and good are at the root of her disposition toward moral agency. Terry credits this moral orientation to a religious upbringing and the devout example her father models. She believes that religious life broadly encourages moral sensitivity and moral reflection. This is supported by her observations of students. Those from families where religion is practiced, irrespective of the particular religion, seem primed to consider and discuss morally salient issues associated with peer relationships, the social dynamics of the class, current events, or the behaviours and motivations of fictional characters.

Terry does not credit her own schooling or the teacher education programs she attended, however, for influencing her moral orientation and sense of moral agency:

> My most memorable teacher was my grade-six teacher at the American International School in Libya. He inspired the children to do really good work, and he always had motivating work for us. He had an expectation that if you're giving in this work, you can do a better job. So if you hand in the work and it's kind of shoddy, not your best, then he'll say, "You need to go back and do a better job at it". He always expected really good work. He was extremely strict in lots of ways. And I think that's what drove me as a student to actually want to do good work, because he expected it and you liked him. You really wanted his approval. He also used to make his own math games. And he was an artist. So he would draw a Christmas or an Easter scene on the window. If you'd done your work, and it's a really good job that you've done, you got a chance to paint the window. So I remember all those things, the wonder and the enjoyment of being a student in his class. He always made it motivating. And he always made it fun. It was so much fun to be in his class. Whenever I think of a teacher that I truly enjoyed, it's him.

The teaching qualities Terry references—motivating students to work hard, maintaining high expectations for student achievement, being creative with pedagogy, and fostering a fun learning environment—while admirable, are not necessarily moral in nature and were not discussed in terms of moral agency. Consequently, they had minimal direct influence on Terry's related practices, although they likely influenced her teaching philosophy and pedagogy more generally.

Further, there were no courses or units of study addressing themes of morality in Terry's undergraduate or graduate education programs. She describes one elective, however, as coming close:

> There was only one course in teacher education that comes close to anything we're talking about, that had anything to do with values. But it was more self-awareness and self-discovery, knowing within yourself things about yourself. It had to do with spirituality, although it doesn't say anything about a particular religion. It more or less discussed that if your core being is soulful and you know yourself well enough to be comfortable with that then you are better able to receive soulfulness in your students. You are better able to see soulfulness in others and help them build up who they are and be comfortable with themselves. You can recognize the needs of students and help them. I just loved that class because it felt like I was being filled up with something important. It comes from who you are as a teacher first, before anything else can happen. But I was the one who chose to be in that course. So in a sense, as a learner you had to be ready to be there.

Thornberg's (2008) empirical work confirms the lack of values education for teachers:

> When I ask them about how they have received or appropriated the values they see as important to mediate to the students, they refer to their own childhood, their personal experiences as children and adults in relation to or interactions with others (their parents, friends, colleagues, and others), and to sources like common sense, personal worldviews, emotions, and personal conceptions. (p. 1793)

Hence, Terry's moral agency is as much a personal ideology as it is a professional endeavour:

> I'm being the teacher I want to be, not just the deliverer of academics. It's not fulfilling any other way. It's not rewarding any other way. No one told me to do that. It is more out of my own interest. . . . Before the character research started, I was already interested in the [students], who they are. And if something isn't quite right, I reflect on it and wonder what's going on. So if there's something amiss I would be going to the student and saying, "What's going on?" Or I'd ask someone, "What is going on in your life?" So I was beginning to build that interest into my practice. But it wasn't specific moral values. It wasn't as developed.

TERRY'S VISION

Terry's intrinsic concern for the moral dimension of teaching, learning, and classroom life crystallized while working with Sean to create the school's Character Development Program:

> When Sean approached me to do the character education research, all of the readings that I did kind of brought everything that I am into focus. And it gave me a way to verbalize. And I just went, "This is so me". I want to instil that in children, as well. The readings I did about character education did give me the language, the philosophy, and a way to articulate who I am and what things are important to me as a person, and why they are important to me. . . . In the end, it just gave me a "Here's what I want. Here's what I want out of them. This is how I want it to look within a classroom or how I want them to work".
>
> I want them to be a better community. I want them to be caring towards each other. I want them to be aware of other people, rather than just themselves. And then I want them to be able to manage themselves

and have self-control. And some of that may not be part of character education, but I'm trying to grow them up more generally, too.

In these passages, Terry expresses a vision for her students and the classroom environment with respect to three mutually dependent moral objectives—the holistic development of each child, a class community, and happiness at school.

Terry assumes direct responsibility for the holistic development of her students. "They are my wards for a whole year, so I'm responsible for how they grow up while they're in my care", she explains. Accordingly, Terry often articulated the catchphrase, "I'm trying to grow them up". She clarifies what this entails with respect to academic, social, and moral development:

> It's all about the entire individual. I want them to not only develop academically but to become better, more aware. This is the age where they are more able to not only think about themselves. I think I can push them to think about other things outside of themselves. . . . If I don't deal with all aspects of the individual, I'm missing a whole part of their development. And I always will question myself, "Why am I not dealing with that part, as well?" It's all connected; so I can't just ignore one part of it.
>
> I want to make sure that if there's something else going on that's not part of their academic life—it could be their social relationships with their peers or something happening at home—I have to make sure that I take the time to listen to them, to make sure that I'm not bypassing all of that stuff in order to just work on their academics. I want to connect with them as a whole child. If things aren't going well outside of school, it shows up in whether they get more easily frustrated, whether they become more impatient, maybe with their friends at recess. And when things go wrong, they'll quickly blow up or react more physically. If I see it, it may just be an anomaly, a bad day. But if it happens again in the near future, I'm thinking there's something else going on here.
>
> The end ideal would be development within themselves and how far they've grown individually, from September to June. . . . Is there some kind of bigger understanding about you by June? I'm hoping to see improvement in the development of who they are as people. Whether it's becoming aware of what they need as learners, let's say, or becoming aware of how to help other people more so than you were before. If you weren't very respectful towards others in September, maybe by June I can see that you are being more respectful.

Terry does not generally distinguish social development from moral development. For her, social development, or the acquisition of pro-social values such as listening and sharing, is underpinned by moral values, such as respect and fairness and, therefore, is essentially moral. Terry does distinguish academic development from social-moral development but usually acknowledges a relationship between them. The relationship was articulated in various ways. On one occasion, Terry suggested a linear relationship, in which social-moral development enables academic development: "Wherever [students] are, socially and emotionally, is connected to how they're going to perform academically". On another occasion, she suggested social-moral development as a higher objective:

I like that they become better readers by the time June comes around. And I like that they can remember mathematical concepts so that they're better prepared for grade-five. That's important, too. But when they become better people, when you see them develop into something that you see they will continue to be as adults, those are the qualities that I'm hoping to develop, to grow in them. Because that's the one that will become more important later on, regardless of what they take at school.

On a third occasion, Terry suggested that moral values connect academic and social-moral development:

There's the whole social growth, which covers some of the respect and compassion, and also integrity and responsibility. And then there's the academic growth, which is more or less responsibility. Where moral and academic growth meet is whether or not the students rely on me and expect me to do it for them. And then that whole confidence thing, too, self-esteem that comes with, "I don't know what to do", or, "Where's this or this?" I'm trying to grow Pia up so she's more confident in what she knows. In the end, her work is always really good, but she doesn't know that at the beginning of a project. So it would be nice for her to see that ahead of time. Heather today was constantly asking for the correct spellings, rather than taking responsibility; "What are the ways I can find out for myself?"

Terry's observations are represented in Stengel and Tom's (2006) five models for the relationship between academic and moral goals—separate, sequential, dominant, transformative, and integrated. A *separate* relationship infers that moral and academic goals are independent of each other and operate in discrete domains. In a *sequential* relationship, either moral or academic goals precede, providing a foundation on which the other may be pursued, as in Terry's first statement. A *dominant* relationship involves consistently prioritizing either moral or academic goals, such that the two are not necessarily connected. This is inferred by Terry's second statement. In a *transformative* relationship, moral and academic goals are linked, whereby one changes, reshapes, frames, or sets an agenda for the other, as with Terry's third statement. Finally, in an *integrated* relationship moral and academic goals are intertwined, mutually supportive, and reciprocally transformative.

Stengel and Tom (2006) further distinguish these five categories, such that when moral and academic goals are separate, sequential, and dominant, the teacher maintains power over students and delivers knowledge as a static entity. When moral and academic goals are transformative and integrated, power is shared between the teacher and students, and knowledge is dynamic and socially constructed. Although Terry articulated positions consistent with sequential and dominant relationships, her pedagogy and interactions with students conform to the latter description and support an integrated relationship overall.

Given the discrepancy between Terry's statements and my perception of her practices, I pressed Terry to clarify her position. She proposed the metaphor of a ladder, with one vertical post representing social-moral development, the other representing academic development, and the horizontal rungs representing moral values that link and support them both:

> The academic helps [students] practice responsibility, and responsibility enables the academic. Then maybe integrity meets up as another bridge. With integrity, it can be integrity with the people and how you treat other people, and integrity in saying to me, "I didn't do my reading last night. I didn't hand it in because . . ." If you're able to tell me that, that's great. Courage might be another bridge—courage in standing up for your friend in a very uncomfortable situation, and then the courage, like Paige, to do public speaking. So there are certain instances where they meet. . . . They support each other because if the social aspect isn't working then they're not as focused. For example, when Heather was having those issues socially it showed up in her math test, and she wasn't focused and couldn't do it. She didn't do well on that test even though she knew the work. . . . Safe is like feeling respected, feeling cared for. And whether or not Paige can trust someone in a social situation is the same as when she has to do something she doesn't like doing academically, but she has to be pushed towards it. Then she feels good that she has had this difficulty and has overcome it.

Terry's description confirms an integrated model, with neither academic development nor social-moral development preceding, dominating, framing, or shaping the other.

The social-moral development of each student is connected to the second objective of Terry's vision, a class community. "It's always you in relation to others. It's always related to how [students] relate to the group", Terry notes. This belief is made concrete by a conception of community grounded in caring relationships among the students:

> In the community, it is important for [students] to be good in their relationships. It's a community that you're trying to grow, and a community means that you take care of each other, not just yourself. And

this is the age where they can start looking to take care of other people. This is where they need to start thinking about other people because it isn't just about them anymore. I would like to see them being a good community, where they're helping each other out and being observant about each other's needs. Doing things to help fix things if things aren't working properly. And you can actually look at other people, and say, "I need to do something here".

This statement suggests an altruistic quality to the relationships and a developmental shift away from the more egoistic position of younger children. The expression of related moral values, including care, helpfulness, responsibility, empathy, compassion, consideration, and sensitivity is implied. It is by these values that judgments and behavioural choices regarding the needs of others are made. As each student matures, socially and morally, he or she is more able to meet these needs, and relationships and community are enhanced. A class community, in turn, provides opportunities and the motivation for students to practice moral reasoning and expressing pro-social and moral values and, thus, increases each student's personal development. The balance of self within community is a recurring theme in Terry's practice, as signalled above in relation to the values messages posted in the classroom and illustrated more thoroughly in the next chapter's discussion on community.

The third objective of Terry's vision is the happiness of each child at school. Terry elaborates:

Happiness is knowing that you've had a great day with your friends and with your teacher. The day just went so well that you felt positive about everything. And I know I want to feel really positive when I come to work. So I want [the students] to look forward to coming to school. . . . And if they were unhappy in some way leaving the school, I think that I would feel sorry for them and would want to help them through that. . . . Going home, there's nothing that's bugging you. Everything has been satisfied emotionally. So at home, before you go to bed, you're not crying about something from school.

For Terry, happiness is an outcome of community: "I think the happiness thing is just building a really close-knit and well-connected community". Relationship values, therefore, are similarly reinforced:

I think they need to be aware if somebody is unhappy. . . . You can feel anything else. Anxiety. You can feel fear. You can feel shy even. But everybody understands or should learn to understand to read people's faces when they are not happy. Can you tell when Alexander is not happy? Can you tell when Paige is not happy? They may say, "Okay, they kind of look uncomfortable". It's that, too. But really, when someone isn't *up* and energized you can see it. You can see it in their body language. And I don't have to explain it, hopefully. People know instinctively when someone is not happy. Like when Frances sulks, you know she's not happy. I'm hoping it becomes easier for them to say to themselves, "That person doesn't look very happy today". . . . Then we should all try and make sure that they leave the school feeling better and positive.

Terry often communicates this expectation in terms of the Golden Rule: "*You* want to be happy coming into school, so make sure that people are happy when *they* come to school. And lend a hand if someone needs it, if someone's feeling a little hurt or down or lonely".

Happiness emerged as an education goal in the 18th century among British and French thinkers (Gilead, 2009). Contemporarily, education philosopher Nel Noddings (2003) asserts, "Happiness and education are, properly, intimately related: happiness should be an aim of education, and a good education should contribute significantly to personal and collective happiness" (p. 1). More particular to moral education, Ancient Greek philosophers claimed that happiness is the outcome and ultimate goal of virtues acquisition (Parry, 2009). Character educators Ryan and Bohlin (1999) attribute a similar observation to George Washington: "There is no truth more thoroughly established than that there exists in the course of Nature, an indissoluble union between virtue and happiness" (p. 222).

It was not a philosophical argument, however, that influenced Terry's vision of happiness. Rather, she was inspired by a fourth-grade schoolteacher in Kanazawa, Japan. By encouraging his students to share their lives, Toshiro Kanamori built a class community characterized by compassion, care, and happiness. His practices are profiled in an award-winning documentary called *Children Full of Life* (Japan Broadcasting Corporation), which aired in Canada (2004) as part of the Canadian Broadcasting Corporation's "Passionate Eye Series". Terry describes how this documentary touched her personally and affected her practice:

> He's one of my role models. I used to watch it every year in September, before I'd go into my new class. He's inspiring, a reminder that this is what it's about. He shows a lot of care for his students in order for them to grow up and be in touch with each other and be in touch with their own feelings. The time he found out that a girl's father had died he was great about dealing with that right out in the open with them. All of that was within the classroom. It's their emotional well-being he took care of. A whole bunch of them went and delivered some stuff to the girl. That didn't even take place in school, but I'm sure it was influenced by what the teacher did and said in class.
>
> I remember other parts that stood out for me. When he was in the classroom and he was talking about their journals. And someone had

said something in their journal about how upset they were because they were being bullied. He waited there, and waited there, and waited there. And not a lot of teachers would be willing to wait that long because there is a comfort level to just waiting. What if he got nothing? Finally, the students started talking about it and started breaking down. But I like that he said, "You guys pretend that you're all nice, but you're hurting somebody else".

And then another time he was really angry about someone wasting time working on that boat project for the pool. And he said, "That's it. You're not going to be doing it with us". What stood out there was what he got from the other kids in the team. They said, "Well that's not fair because the punishment didn't fit the crime". And the kid who said that was scared to death to say it to the teacher. He was shaking. But he had the courage to say it anyway. And in the end, the teacher said, "That's good". He acknowledged the courage, that the kid had made a good argument and had shown growth. The growth that they get out of being able to think through something and to voice an opinion about it, in a respectful way, or to stick up for somebody because they're showing compassion. It's a lot of growth in being able to think it through, to voice it, to have the courage to voice it and then see the results of democracy and how the world works. It's a huge learning lesson. It's a good life lesson. I loved that part.

And he was the one who said [students] need to come to school and be happy. "Make sure that you come into school happy and other people are coming into school happy. And if they are not happy then figure out why not. Is there something that you can do?" And that's where I got that. It's so simple and it's easy for them to understand. And then that courageous kid in an interview was asked, "Why did you do that? Why did you stick up for him?" He said, "Well the teacher says that everyone has to come to school happy, and we have to be happy at school".

Terry credits Kanamori's practice for particularly influencing the objective of happiness. Yet, her recollections also reference the other two objectives of her vision. Social-moral development of students is implied when Terry says, "He acknowledged the courage, that the kid had made a good argument and had shown growth". Community is implied when Terry notes that the teacher's care helps students to "be in touch with each other". Although presented as three objectives, Terry's vision is essentially singular, assimilating mutually beneficial ideals of student growth, development, and well-being, with positive classroom relationships.

Hammerness (2002) claims that school context is critical in enabling teachers' vision. When a vision coincides or aligns with the school's resources, collegial environment, administrative philosophy, and policies, for example, it is more readily achievable. Terry recognizes this to be the case at Middlevale:

> The type of school that I'm in fits very well with my expectations of what a whole individual should be, that this place is about educating more than just the academics. The school is a school that talks about educating the whole child, rather than just their mental aspects. And we do talk about confidence. We do talk about self-esteem and respect for others. But because we are a school that has the small classes, as well, it's sort of built in that the expectation is you have to be aware of others and be kinder and more compassionate to others.

Given the limited disparity between Terry's contextual reality and her vision, the vision serves as a source of energy, hope, motivation, and inspiration (Hammerness, 2002).

Even still, Terry occasionally experiences challenges while trying to achieve her vision. Despite abundant justification in education literature, a supportive school context, and the consistent endorsement and encouragement of administrators, Terry struggles, conceptually, with the place of moral education in the classroom:

> And I know it's not my main role. It's obviously their family's role to instil those values in them. I'm just on the periphery. . . . I don't know how much a family would be comfortable with values-based education. It is the family's responsibility, really, to instil values. I don't want to go into new territory of doing that. However, if the values are universal enough and talked about enough in very explicit ways, I think it's safer for me to talk about with them.

These doubts give Terry a reason for pause and reflection but are not a deterrent. She continues to openly and explicitly teach morality, with a few self-imposed restrictions that have been determined through experience and instinct:

> There are issues that I will shy away from. There are issues that I'm not comfortable with. For example, [the holocaust novel] *Number the Stars*. There are issues in that book that relate to genocide and hatred that I did not touch with these kids because they didn't bring them up. It's hard for them to understand a lot of the grey area. I tend to shut them down with anything that's too difficult for their age.
>
> Anything to do with religion—I would be uncomfortable with that because it's not my place. All I would say is, "Everybody has a right to their beliefs and to follow which religion they wish to follow". It's not my place to say anything more than that because it's not up to me

to put forth any kind of religious ideas. But maybe to just make them aware that all religions have expectations of how a good person should behave. The other thing I would be uncomfortable with is the abortion issue.

These restrictions that Terry identifies are topics of discussion, rather than moral values or principles. This is an important distinction. Terry readily discusses with students universal and objective ideas of respect, acceptance, fairness, and compassion, for example, but intentionally avoids particular contexts and applications, including genocide, religion, and abortion.

Additionally, achieving this vision necessitates the students' commitment, cooperation, and partnership. When the students are unwilling or unable to support the objectives, the gap between vision and classroom reality cannot be closed. Terry recalls the following occasion:

Last year was not a great year. I kept trying to teach them to be more reflective ahead of time, before anything bad happens. I used the values all the time. But it wasn't happening. Am I going to see results? I'm not going to see results. That was so frustrating for me. . . . I didn't remain as patient as I could have been, and I got angry a lot. It was not an admirable year at all. There were moments that were really, really good, but in general, if you look at the whole year, I would say that it was a thumbs-down year. I don't feel good about it, but it taught me a lot about what I could handle and when I needed help.

So last year was a year when I learned to actually not handle it myself and just let it go and give it to somebody else, because it's just over-whelming. I admit it. I was pushed to the point where I was very stern with them, and I lost my temper with them. And it wasn't fun at all. And I just sort of had to let it go and say, "I can't". It isn't healthy for either party, the class or myself, to continue rehashing the same arguments again and again. It was not going anywhere. So someone else needs to hear it because then at least both parties get a different perspective.

I decided to let Jonathan handle the boys from a discipline perspective and to let Tom [the school psychologist] help them more from an indi-vidual, behavioural perspective. I just couldn't deal with it. I was talking back at them when they were arguing with me. And I didn't like that. I was very upset and frustrated and angry. And I couldn't get through to them. It felt like I was hitting my head against a brick wall. And then, emotionally, I was not available for the ones who needed me and who were more positive. So I just handed it over.

In doing so, Terry prioritized her own happiness, not as an end but as a means for ensuring the happiness and well-being of the other students.

Although Terry was grateful for Middlevale's resources, she experienced disappointment and a sense of personal failure in having to forgo her vision.

Terry was challenged to achieve her vision on another occasion by an administrative decision she believed undermined the growth and development of a particular child:

> I had a kid in grade-four who was incredibly anxious. His mother was here the whole day, otherwise he wouldn't come to school. He didn't do much academically. The expectations had to be lowered for him. I don't like doing that. And then he wouldn't show up for school and missed a whole bunch of days. At that point, I couldn't work with him because it wasn't just about him. It was also about the mother, who wouldn't leave. He knew that he could go to her anytime he wanted if he had a hard time. It doesn't really give him any kind of coping strategies. But if admin allows the mother to come, I can't do anything about it. When it came time for the field trip, everyone knew that behaviour had to be good in order to earn the trip. He hadn't earned the trip in my mind because he could do whatever he wanted. And he didn't even have to show up for school. He was still allowed to go on the trip. So I washed my hands of it. I thought that we could have done more to guide the separation from home and help his moral and social development. But I couldn't help him grow up without more support from admin. So I put a limit on where I'll go with morality. When I start to feel that it's draining me and I have to save myself, then I don't go there.

This incident coincides with those described by Campbell (1997) regarding administrative decisions that contradict what teachers believe to be right or good. According to Campbell, teachers suspend morality when they do not take a stand against such decisions. To suggest that Terry suspends morality, however, is misleading and disingenuous. Her struggle with ethical dilemmas that might compromise her vision resonates in the following passage:

> When I pass on the students I feel like I'm letting them down because I'm not putting in the effort as much as I should be or could be. I feel badly because it's not what I want to do, and it's not what I should be doing, but in order to stay calmer and to be more available to others I have to do it. I don't feel good about it. I know that when I look at them in grade-five, I could have done more for the ones who were in trouble. I could have done more to grow them up.

Several ethical perspectives are, in fact, revealed in how Terry comes to terms with a decision to "pass on the students". In noting her own well-being, Terry reflects an ethical egoist perspective (Shaver, 2010):

> By February, I was going home and I was crying. I was so tired. And at some point I just had to sort of break off. And I just had to say to

myself, in order to conserve my sanity and my energy I can't put in that much energy into fixing them myself.

A utilitarian perspective, to provide benefit for most, is noted in her desire "to be more available to others". Consequentialism, more generally, is reflected in the statement, "when I look at some of them in grade-five". Care ethics, as related to Terry's personal relationships with students, is hinted at in the statement, "I feel like I'm letting them down"; and deontology, as duty and responsibility, is expressed when she says, "what I should be doing". This demonstrates the complexity of Terry's moral reasoning and the multidimensional ethical framework with which she processes issues of a moral nature. There is no doubt that morality was never suspended, although Terry's vision might have been.

EMERGENCE OF MORAL AGENCY

In the first week of school, Terry's practices did not disclose this complex moral vision to which she is so committed. Terry was gentle, polite, and pleasant to the students; friendly and helpful to staff and teachers; and generous and hospitable to me. The students appeared to be happy and were cooperative and well behaved. The classroom environment was lively, relaxed, and positive. Yet, there was little by way of overt moral instruction, and Terry rarely used moral language or framed messages in terms of right and wrong. She told a group of girls working together in the hallway, for example, "Do not disturb anyone please", but did not reference showing sensitivity or having consideration for others. She instructed Zeth, "Keep your tie done up while you are here. You can loosen it a little if you like, but you need to look respectable". She did not extrapolate "looking respectable" to demonstrating respect and responsibility. While introducing the learning buddies program, Terry coached her students, saying, "The grade-twos look up to you, like you looked up to your buddy", but did not relate this to being moral role models. These and other similar comments and directives were expressed as school conventions, with no indication that students understood them as anything more than rules to follow. This was intentional, in order for Terry to quickly and easily establish schedules, routines, procedures, and behavioural and academic expectations related to non-moral values of efficiency, safety, timeliness, orderliness, tidiness, manners, listening, sharing, and group work. Terry was, nonetheless, highly aware of the moral underpinnings relevant to each situation. She explains:

> Everything that you do, even how you line up to leave the class, is a reflection of whether or not you are respectful of someone waiting for you, especially when you're just talking. And then the same when coming in from recess and being ready to start. Are you being respectful of the time that somebody's giving to you?

The moral core was increasingly made explicit as the first month of school progressed. By mid-October, Terry's moral agency had fully emerged, in pursuit of her vision. No longer accepting mere obedience and conformity to norms and conventions, Terry encouraged the students to reflect on and understand their behaviours, assumptions, and attitudes through a moral lens and to enact behavioural choices and decisions that are good and right, not only for themselves but also for each other and the class as a whole. A range of relevant practices was revealed, extending across conceptual and empirical literature and sampling applied aspects of character education, care ethics, and cognitive development. A flippant and spontaneous thought crossed my mind in an "ah-ha" moment as I browsed the professional library behind Terry's desk: *You may like Lickona, but you seem more like Coloroso to me.* Terry previously noted that, while researching moral education, she had been particularly inspired by Lickona's work:

> I believe that we have to make sure we're good people and becoming better all around. I'm working on the students' values. But how they act upon those values is their responsibility to the classroom community and to the outside community, as well. That means everything from the classroom to home life and then local and global communities. But it starts with making sure that you have the right values. So when I read Lickona, I thought, this is it. This is what I do.

Although both authors are represented on her shelves—Thomas Lickona (1991, 2004) associated with character education, and Barbara Coloroso (2005) associated with care ethics—Terry admitted to never having read Coloroso's work and was not familiar with care ethics theory or the academic work of its major proponents, Carol Gilligan and Nel Noddings. Yet, she intuitively redirected her professional life toward teaching in accordance with an ethic of care: "I was attracted to the connection that you can have with the students". In addition, Terry echoes Noddings (2002, 2010) when identifying her relationships with students as an opportunity for moral education: "I still want to make a good connection to my students and get to know them really, really well, so that there's a respectful relationship and a trust that I'm going to help them develop". Terry was not familiar with the theoretical body of literature related to cognitive development, either. She had, nonetheless, collected several practical books on conflict resolution and peer mediation and expressed goals related to moral reasoning:

> I say to them, "It is the year where I'm going to ask you to make decisions more and more on your own. So this is one of those times where you have to make a good decision on what is the right thing to do. You have to make a choice on how you will react or behave. And you have to say to yourself, 'What is a good choice, not only for myself but for others involved, as well?'" So a lot of conversations will just keep

happening, based on this theme. And I try to help them learn how to make these decisions.

Although helping me to detect Terry's moral education practices, character education, care ethics, and cognitive development do not determine the conceptual framework by which they are represented and discussed in this book, mainly because Terry does not approach moral education from a theoretical or philosophical perspective. Rather, seven practices emerged in her efforts to achieve the moral objectives of her vision: (a) expressing moral values in the teacher's personal and professional conduct, behaviours, and practices; (b) nurturing a class community; (c) fostering caring relationships with students; (d) providing virtues instruction; (e) encouraging and facilitating morally laden discussions; (f) disciplining with a goal of self-regulation; and (g) implementing service activities. These practices and related strategies are explored in the next two chapters as teaching morally and teaching morality (Fenstermacher, Osguthorpe, & Sanger, 2009). Chapter Four addresses how Terry teaches *morally*, or in morally justifiable ways, as a desirable end. This includes the first three practices of expressing moral values in one's own conduct, behaviours, and practices; nurturing a class community; and fostering caring relationships with students. Chapter Five addresses how Terry teaches *morality* by describing more direct and deliberate practices for imparting a moral education. This includes the remaining four practices of providing virtues instruction; encouraging and facilitating morally laden discussions; disciplining with a goal of self-regulation; and implementing service activities. Both chapters acknowledge a synergy between teaching morally and teaching morality, such that the empirical distinction is sometimes difficult to rationalize and maintain.

REFERENCES

Biesta, G.J.J & Tedder, M. (2006). *How is agency possible? Towards an ecological understanding of agency-as-achievement* (Working paper 5). Exeter: Learning Lives Project. Retrieved May 14, 2010, from http://www.learninglives.org/papers/working_papers/Working_paper_5_Exeter_Feb_06.pdf.

Billig, S.H. (2000). The effects of service learning. *School Administrator, 57*(7), 14–18. Retrieved May 15, 2012, from http://proquest.com.

Billig, S. H. (2009). It's their serve. *Leadership for Student Activities, 37*(8), 8–13. Retrieved May 15, 2012, from http://proquest.com.

Campbell, E. (1997). Administrators' decisions and teachers' ethical dilemmas: Implications for moral education. *Leading & Managing, 3*(4), 245–257.

Campbell, E. (2003). *The ethical teacher*. Maidenhead: Open University Press McGraw-Hill.

Campbell, E. (2004). Ethical bases of moral agency in teaching. *Teachers and Teaching: Theory and Practice, 10*(4), 409–428.

Campbell, E., & Thiessen, D. (2000, April). *Moral and ethical exchanges in classrooms: Preliminary findings*. Paper presented at the Annual Meeting of the American Educational Research Association, New Orleans, LA.

Campbell, E., & Thiessen, D. (2001, April). *Perspectives on the ethical bases of moral agency in teaching*. Paper presented at the Annual Meeting of the American Educational Research Association, Seattle, WA.

Coloroso, B. (2005). *Just because it's not wrong doesn't make it right: From toddlers to teens, teaching kids to think and act ethically*. Toronto: Viking Canada.

Fenstermacher, G. D. (2001). On the concept of manner and its visibility in teaching practice. *Journal of Curriculum Studies, 33*(6), 639–653. doi: 10.1080/00220270110049886.

Fenstermacher, G. D., Osguthorpe, R. D., & Sanger, M. N. (2009, Summer). Teaching morally and teaching morality. *Teacher Education Quarterly, 36*(3), 7–19. Retrieved February 25, 2010, from http://vnweb.hwwilsonweb.com.

Furco, A., & Root, S. (2010, February). Research demonstrates the value of service learning. *The Phi Delta Kappan, 91*(5), 16–20. Retrieved May 22, 2012, from www.jstor.org.

Gilead, T. (2009). Progress or stability? An historical approach to a central question for moral education. *Journal of Moral Education, 38*(1), 93–107. doi: 10.1080/03057240802399483.

Goodman, J. F. (2006). School discipline in moral disarray. *Journal of Moral Education, 35*(2), 213–230.

Hammerness, K. (2002). *Learning to hope, or hoping to learn? The role of vision in the early professional lives of teachers*. Paper presented at the annual meeting of the American Educational Research Association, New Orleans, LA, in April 2000. Retrieved August 14, 2010, from http://ezproxy.qa.proquest.com/docview/62182843?accountid=14771.

Hart, D., Matsuba, M. K., & Atkins, R. (2008). The moral and civic effects of learning to serve. In L. P. Nucci & D. Narváez (Eds.), *Handbook of moral and character education* (pp. 484–499). New York: Routledge.

Howard, R. W. (2005). Preparing moral educators in an era of standards-based reform. *Teacher Education Quarterly, 32*(4), 43–58. Retrieved February 26, 2010, from http://proquest.umi.com.

Howard, R. W., Berkowitz, M. W., & Schaeffer, E. F. (2004). Politics of character education. *Educational Policy, 18*(1), 188–215. doi: 10.1177/0895904803260031.

Jackson, P., Boostrom, R., & Hansen D. (1993). *The moral life of schools*. San Francisco: Jossey-Bass.

Johnson, D. W., & Johnson, R. T. (2008). Social interdependence, moral character and moral education. In L. P. Nucci & D. Narváez (Eds.), *Handbook of moral and character education* (pp. 204–229). New York: Routledge.

Lickona, T. (1991). *Education for character: How our schools can teach respect and responsibility*. New York: Bantam Books.

Lickona, T. (2004). *Character matters: How to help our children develop good judgement, integrity, and other essential virtues*. New York: Touchstone.

Morgan, W., & Streb, M. (2001). Building citizenship: How student voice in service-learning develops civic values. *Social Science Quarterly, 82*(1), 154–169. doi: 10.1111/0038–4941.00014.

Noddings, N. (2002). *Educating moral people: A caring alternative to character education*. New York: Teachers College Press.

Noddings, N. (2003). *Happiness and education*. Cambridge: Cambridge University Press.

Noddings, N. (2010). Moral education and caring. *Theory and Research in Education, 8*(2), 145–151. doi: 10.1177/1477878510368617.

Parry, R. (2009, Fall). Ancient ethical theory. In E. N. Zalta (Ed.), *The Stanford encyclopedia of philosophy*. Retrieved May 18, 2011, from http://plato.stanford.edu/archives/fall2009/entries/ethics-ancient/.

Richardson, V., & Fenstermacher, G. D. (2000). *The Manner in Teaching Project.* Retrieved July 29, 2008, from www.personal.umich.edu/~gfenster.

Richardson, V., & Fenstermacher, G. D. (2001). Manner in teaching: The study in four parts. *The Journal of Curriculum Studies, 33*(6), 631–637.

Ryan, K. (1993) Mining the values in the curriculum. *Educational Leadership, 51*(3), 16–18. Retrieved April 16, 2011, from http://ezproxy.qa.proquest.com/docview/62786313?accountid=14771.

Ryan, K., & Bohlin, K. E. (1999). *Building character in schools: Practical ways to bring moral instruction to life.* San Francisco: Jossey-Bass.

Salls, H. S. (2007). *Character education: Transforming values into virtue.* Lanham, MD: University Press of America.

Schuitema, J., ten Dam, G., & Veugelers, W. (2008). Teaching strategies for moral education: A review. *Journal of Curriculum Studies, 40*(1), 69–89. doi: 10.1080/00220270701294210.

Shaver, R. (2010, October). Egoism. In E. N. Zalta (Ed.), *The Stanford encyclopedia of philosophy.* Retrieved Dec. 4, 2010, from http://plato.stanford.edu/archives/win2010/entries/egoism/.

Stengel, B. S., & Tom, A. R. (2006). *Moral matters: Five ways to develop the moral life of schools.* New York: Teachers College Press.

Strike, K. A. (2008). School, community and moral education. In L. P. Nucci & D. Narváez (Eds.), *Handbook of moral and character education* (pp. 117–133). New York: Routledge.

Thornberg, R. (2008). The lack of professional knowledge in values education. *Teaching and Teacher Education, 24*(7), 1791–1798. doi:10.1016/j.tate.2008.04.004.

Watson, M. (2003). *Learning to trust: Transforming difficult elementary classrooms through developmental discipline.* San Francisco: Jossey-Bass.

Watson, M. (2008). Developmental discipline and moral education. In L. P. Nucci & D. Narváez (Eds.), *Handbook of moral and character education* (pp. 175–203). New York: Routledge.

4 Teaching Morally
Practices That Are Good and Right

Fenstermacher, Osguthorpe, and Sanger (2009) define teaching morally as follows:

> To teach morally is to teach in a manner that accords with notions of what is good or right. That is, to conduct oneself in a way that has moral value . . . the teacher is being a good or righteous person. (p. 8)

This definition corresponds with Campbell's (2003) description of the teacher as a moral person: "The teacher who strives to empathize with students and colleagues, who aims to be fair, careful, trustworthy, responsible, honest, and courageous in the professional role" (p. 23). Both characterizations are philosophically rooted in a virtue ethics conception of morality. I expand the notion of teaching morally beyond the teacher's character dispositions, however, to also include the pursuit of morally desirable and justifiable goals or ends. This chapter develops three of Terry's related practices—expressing moral values in the teacher's personal and professional conduct, behaviours, and practices; nurturing a class community; and fostering caring relationships with students. A morally educative potential is embedded in each, and Terry is intentional in pursuing it. Although teaching morality is the subject of the next chapter, I address it in relation to these practices, as an outcome of teaching morally, rather than a discrete practice for teaching morality. As the portrait progresses, however, the distinction between teaching morally and teaching morality becomes more difficult to maintain.

This chapter is arranged in three sections, one for each practice. The first section identifies and illustrates many moral values expressed in Terry's conduct, behaviours, and practices. In accordance with Campbell's (2003, 2004; Campbell & Thiessen, 2000, 2001) empirical and conceptual work, moral values of fairness, respect, kindness, and honesty are used to organize the discussion. I also illustrate how Terry models morality for students, as a means of moral instruction. The second section describes Terry's efforts to nurture and sustain a morally infused class community. Again, I note the means by which Terry utilizes this environment to impart lessons and messages of a moral nature, particularly related to helpfulness and responsibility. The final section recounts how Terry fosters caring relationships with each student and how these relationships provide a model for the relationships she hopes to foster among students. Despite being treated independently, these three practices interact in mutually supportive ways, as will be evident by the end of the chapter.

EXPRESSING MORAL VALUES

Fairness, respect, kindness, and honesty, as moral values, are expressed in association with many others. Campbell (2003) notes that fairness is rooted in the principle of justice and is related to both equity and equality. In education discourse, equity refers to pursuing the same desired outcomes for every student. Consideration is given to individually accommodating students' needs and exceptionalities. Different resources, experiences, and opportunities are provided, and different restrictions and boundaries may be applied. In this regard, fairness is understood as predominantly consequentialist. Equality, however, may be viewed as predominantly non-consequentialist, prioritizing consistency and constancy over outcomes. The same resources, experiences, and opportunities are objectively and impartially offered to every student, and the same restrictions and boundaries are objectively and impartially applied. Respect is often associated with fairness, courtesy, care, kindness, trust, consideration, sensitivity, and thoughtfulness (Campbell, 2003; Lickona & Davidson, 2005). Identifying these additional moral values helps to distinguish authentic expressions of respect from non-moral expressions of compliance, subservience, awe, politeness, good manners, conceit, and self–aggrandizement (Campbell, 2003). For example, respect is expressed when compliance with school policy is underpinned by courtesy, consideration, and responsibility and when politeness and manners are underpinned by care, kindness, sensitivity, and thoughtfulness. Similarly, moral kindness is a visible manifestation of care, sensitivity, empathy, compassion, consideration, and thoughtfulness. It is distinguished from behaviours and actions that appear kind, or are nice, by association with other moral values. For example, a teacher may greet students at the classroom door every morning, according to school policy. This is not an example of kindness, unless also motivated by caring and thoughtful intentions. Finally, honesty is an aspect of integrity related to being truthful about oneself and being truthful to others. The former, as self-honesty, requires self-awareness and is, thus, the opposite of self-deception. Being truthful to others is defined as accurate and authentic self-representation and avoiding the telling of lies. Accordingly, honesty may be associated with humility, genuineness, and sincerity. This range of moral values is profoundly evident in Terry's conduct, behaviours, and practices.

Fairness

Terry considers fairness to be a desirable character trait, a self-evident foundation of teaching practice, and a necessarily ubiquitous attribute of classroom life. This became apparent on the first day of school, when Paige recounted a rumour labelling Terry as a strict teacher. Terry laughed and replied, "I hope I'm also fair and nice about it". Terry's aspiration to be fair often manifests in equitable practices, justified by a deeply held belief in human equality and an unconditional love for all children. As both a professional and a moral obligation, Terry acknowledges and supports each students' peculiarities and exceptionalities by way of ensuring all students' psychological and physical well-being, individual sense of worth, and academic success. Several moral values are indicated, including care, compassion, sensitivity, empathy, and thoughtfulness. This conception of fairness does not create for Terry conflicts or dilemmas between equity and equality, as other teachers have experienced (Campbell, 2003; Colnerud, 1997; Fallona, 2000), and, in fact, enables Terry to help 9- and 10-year-old children understand and accept different treatment as being fair.

Terry's practices for ensuring the psychological and physical well-being of students predominantly reflected a principle of fairness as equality. Three incidents illustrate this. Two relate to students' psychological well-being. In the first, a few students were snacking at their desks, mid-afternoon. Terry announced, "This doesn't feel good to me because the rest of us don't have anything to eat right now". She sensitively assumed that other students would also feel badly about the behaviour and requested that the class eat together during designated or agreed upon times. On another occasion, Terry was asked to mediate a disagreement between four girls who were sharing cookies at their table and the remaining 11 students who were being excluded. Despite that this had occurred during lunch and everyone was eating, the behaviour was deemed unfair because it needlessly divided the class. The third example is recounted anecdotally and relates to students' physical well-being:

> Kathy progressed slowly up the stairs from recess, cradling her left arm in her right, tears staining her cheeks. Bonnie accompanied her, with an arm protectively around Kathy's shoulders and a face set with purpose. The two girls crossed the classroom threshold and walked the straight line to Terry's desk. Terry instantly judged that this was not to be downplayed. Tears were unusual for Kathy, and such determination was unusual for Bonnie. Terry rose from her chair and approached the girls around the front of her desk, crouching to eye-level with Kathy. Not looking at Bonnie, she said, "Thank you, Bonnie. If you were not directly involved in this, you can go and sit at your desk now". Bonnie did, but reluctantly, closely watching the proceedings from there. "Tell me what happened, Kathy", Terry continued.

"Noah hurt my arm", she replied with an unsteady voice.

Noah had entered the room quietly, along with the rest of the class. Without drawing attention to himself, he was sitting at his desk appearing to read a novel. "Noah, please join us", Terry requested, only briefly glancing his way. Noah took his time, moving in uncharacteristic slow motion. He retrieved the bookmark from his desk, and returned it to its spot in the book. He closed the book and placed it inside his desk cubby. He stood and pushed in his chair. Then he sauntered toward Terry's desk from the back of the classroom, never once looking up. All eyes in the room had followed these precise movements. Noting this, Terry reminded the class, "The rest of you can continue with your reading". Eyes dropped into the pages of various books.

Terry began. "So, I'm going to hear what happened from each of you, one at a time. I want you to listen to each other, as well. After we've heard both accounts, we'll talk". Terry listened without interruption as first Kathy then Noah told their versions of the incident. Although the other students kept their eyes in their books, the unusual silence and stillness of the room was evidence that they were intently listening, as well. No pages turned. This didn't concern Terry because the situation was already public.

When both had finished, Terry said, "There are two sides to this story, two versions, so be careful what you say. Would either of you like to add anything, now that you have heard the other version?" Kathy and Noah remained silent. "It seems to me then that this is the result of a misunderstanding. Can we agree to this?" Kathy, still sniffling, was the first to nod. Noah agreed a little more reluctantly. Terry continued, "Kathy, you must learn to know the difference between helping somebody and taking their ball. Noah, do you understand why grabbing Kathy's arm was not right? A physical reaction to this is inappropriate and cannot happen, ever". While I strained a bit to hear most of this dialogue from my perch at the back corner, the last statement was remarkably clear.

Kathy and Noah returned to their seats and took out their books. Terry let the quiet reading continue for a few moments longer than usual, allowing each child the privacy of his or her thoughts. Although appearing busy at her desk, Terry seemed lost in her own thoughts, as was I. This situation had some layers. Noah's anger at Kathy was justified, and Terry recognized this and was sensitive to it. But she would not accept his physical response, intentionally addressing that for everyone as a categorical imperative.

Ensuring each child's sense of worth involved applying equal and equitable practices in pursuit of three goals: all students have a *voice* in the classroom, all students are given Terry's personal attention, and all students participate in classroom life. Terry explains her goal of student voice in terms of fairness as equality, but hints at equity: "Fairness with respect to having your opinions voiced, even though maybe you're too timid to raise your hand. But at least to have a say in the class during the day so we can hear you". Having a say was broadly and symbolically interpreted to account for the variety of media through which students may express and assert themselves. For example, equality was evident when Terry posted everyone's assignments on the bulletin board, clothesline, door, or in the hallway. This was logistically possible because of the small class size. Yet, Terry did not consider posting some and not others. Further, in preparing several bulletin boards for her, I was never instructed on a particular orientation, despite obvious variations in the quality of students' work. Rather, the pieces were randomly and impartially arranged.

During lessons, Terry similarly aimed for participation from all students. She encourages this with the following statements and questions: "Hold on. You've talked twice. Who hasn't spoken?"; "Has everybody spoken who wants to?"; "Let's hear from someone who hasn't spoken all day"; and "I'd like someone to read. Who haven't I heard from today?" Terry might also target a particular student, asking, for example, "Do you have a comment on this?" She explains:

> There are some kids who never raise their hand or do not raise their hand often enough, and I don't hear from them. So I kind of need to hear from them and call them out. Sometimes it's a matter of hearing from them and giving them a voice. And sometimes it's just a matter of keeping them on track with respect to reading, especially.

Terry's last remark reveals this practice to have both moral and pedagogical intentions and, thus, reinforces the integration of her moral and academic objectives, as discussed in Chapter Three. Further, pausing after asking a question and before accepting an answer, Terry equalized opportunities for slow and fast cognitive processors to participate and prevented students whose hands are often raised first from dominating lessons. She signalled this intention by making statements such as the following: "Don't say it out loud so other people get a chance to think about it first"; "Hands down for a second. I'm just going to watch all the pens till everyone is done"; and "Please let him do his thinking because you've already done yours". In a particular example, Bonnie was in the process of writing a math answer on the whiteboard when Connor called out, "You missed some".

Terry swiftly replied, "It's okay. Let her do it her way, and then we'll see what we think after". When Bonnie finished, Terry turned to Connor, saying, "Connor, you have a different opinion about what is on the board. Please come up with a different coloured marker and show us". Terry, thus, enabled both voices to be heard.

While all students are expected to participate during lessons, they may do so by various means, according to their desires, interests, comfort levels,

and abilities. This includes answering or asking a question, suggesting a strategy for generating an answer, clarifying or offering a comment, providing feedback, making a suggestion, and reading aloud. Further, when students alternate reading aloud, in the order of their seating arrangement, they all have a voice in the activity. Yet, Terry allows the stronger readers to read longer passages, simply because it takes less effort for them to do so. In addition, she provides corrections in pronunciation for stronger readers more frequently than for weaker readers and allows stronger readers to struggle longer and harder with more advanced words. Justifying this practice, in relation to Alexander, Terry reveals the sensitivity, thoughtfulness, and care by which it is mediated:

> Alexander wanted to read a page in the story to the class. He knows that reading is not his strength. But I said, "Great, I'm so glad you want to read". And then he was having such difficulty reading that page that I know if I stopped him at every single mistake he wouldn't ever raise his hand again to read for a long time.

Terry's desire to personally attend to every student is also understood as fairness, and is enacted both equally and equitably. She explains:

> Fairness, in the sense that I should be able to give them equal treatment, equal attention at some point. If it doesn't happen during the day then somehow throughout the week I've got to make up for the attention that I have not been paying them. I'm more concerned with making sure that they all feel like they're being cared for, that they're all getting attention from me, and that I'm not forgetting about any of them. . . . Because I don't want them going out of school and feeling like they were forgotten or that they were ignored. If I was a student and my teacher didn't talk to me, I would think that my teacher doesn't really care about me.

By way of example, Terry recounted the following situation:

> Paige and Frances were alone at their desk. Noah wasn't here, and Zeth had moved over here to sit with some boys. . . . I just saw this as an opportunity to be with them. I can enjoy their company. And it might help make the other girls realize that they can, too. I wanted to give them a little bit of attention. I just felt that I needed to do that.

The importance Terry places on giving all students her attention is reinforced by the regret she feels when she is unable to do so:

> I really wanted to have a student conference at the end of the year with every single student, to catch up with them for five minutes, see how the

year went and what their thoughts and goals are for next year. I didn't want to let them go and not give them that little message that they can carry forward. I like one-to-one with my kids where it's just honest and heart-to-heart stuff. Zeth had a great conference with me. I caught up with Sammie the last day. I had a conference with Heather and Noah, too. But that was only four of the 15 kids. I just ran out of time. I would have liked to have this conference with all of them.

As with voice, the nature of Terry's attention varied to serve students' particular needs. Bonnie required attention while doing seatwork in order to progress; Kathy was affectionate and received many hugs; Pia required help to start projects but once underway worked independently; Zeth required one-on-one coaching prior to presentations; and Frances received emotional support and encouragement to manage anxiety. The students did not challenge as unfair Terry's equitable practices related to voice and attention. For voice, the practices may have been understood as standard pedagogy. This is less likely to be the case with respect to giving personal attention, however. A broader context of equality seems to have been established in which any such challenges would be trivial, perhaps even insensitive.

Finally, participation in classroom life is an expectation Terry equally upholds for all students to ensure their sense of worth in the classroom. Yet, the absence of a duty wheel, or other system for allocating daily tasks, enabled students to participate according to particular desires, interests, comfort levels, and abilities. Bonnie routinely circulated books, worksheets, and other materials in preparation for a lesson; Sammie regularly organized and tidied the classroom shelves; Kathy often helped others to clean their desks and lockers; and Paige maintained the whiteboard, updating it every morning by changing the date and school day and erasing irrelevant information. By the third term, Paige was Terry's uncontested scribe, also recording daily messages and homework assignments. Terry intentionally reinforces a system of volunteering. As with other equitable practices, her justification resounds with empathy, sensitivity, and thoughtfulness:

> Some kids are more comfortable going out and checking, for example, with the grade-five teachers whether or not they are using the laptops. I know some kids won't want to go to that end of the hallway because it's the older kids. Sometimes I do make them do it, but I get them to go with a buddy. Then at least they are doing it in a scaffolded manner. They're still doing it but with support. You can see by their faces. If I say, "Alexander, would you mind going to the grade-five classroom and checking to see whether or not they have the laptops?" If his face isn't eager and he's hesitant, then I say, "Would you like to go with Connor?" . . . When a student takes another student over to the office, I always need two people to do that. There are some kids who just don't care to do that.

Terry's practices for ensuring the academic success of all students predominantly reflected a principle of fairness as equity. She illustrates this in the context of a math test:

> In math tests, I walk around and help some kids because I know that they know the work. In a test situation they're not always using what they know. And I know it's got to be there because they've done problems that are similar to it. So I want to make sure that they show me what they really know, instead of shutting down or not remembering because it's a test. Sometimes in a test situation it doesn't feel the same, and they get a little bit stressed. I just want to be able to make sure that they have the opportunity to show me what they really know, instead of making mistakes when I know they know better. It gives them more confidence.

A pedagogical argument might criticize this practice as unfair because it is unequal. A moral argument, however, recognizes the underpinnings of compassion, care, sensitivity, and thoughtfulness and judges this as equitably fair. Many such accommodations were routinely provided, either at Terry's discretion or at the request of students. Also regarding tests, for example, Terry occasionally said to some students: "Wait a minute. I'm going to check your work to make sure it's the best you can do. Stay here because I might give it back to you to do more". In another example, Terry noticed that Zeth became very anxious for in-class and school-wide presentations, such as book talks, skits, current events reports, musical recitals, and dramatic performances. I was asked on several occasions to privately coach him on being louder, going slower, periodically looking up from the page, and standing straight and still. More importantly, though, my role was to boost his confidence. Terry also solicited the help of Tom Sinclair, the school psychologist, to provide more general coping strategies. Zeth was never excused from this academic expectation, however. Further, Alexander received accommodations for difficulties he was encountering in the language arts program:

> Before the language arts tests, Alexander goes into the hall to review the vocabulary. And then he can decide when he is ready and come back in and join the class to write the test. I don't consult anyone on this accommodation. I just know that he would be able to succeed better if he had more time—instead of being pressured with time. It's a confidence thing. Because if he's pressured to do it when he's not ready then he won't show much at all. And then he won't feel so great. He's just as capable as everyone else, but he needs a little extra time. I'm not going to penalize him if he needs extra time. So if he's going to succeed using other tools or extra time, then by all means.

Terry is also willing to extend this accommodation to others:

> And I want the others to know that they can have the same thing, too, if they feel they need it. By all means, ask for it, and you can have the same thing. Just so you can show me, for real, what you can do.

The students do, in fact, advocate for themselves and negotiate individual accommodations. The following exchange recounts one such instance.

TERRY: Alternative math groups are to make sure each of you gets what you need. It doesn't indicate your overall ability in math. If you already know how to do something then you don't have to relearn it. You will do something else. I don't want anyone feeling badly about not being in an alternative group. The next unit might be something different, and different students will be in the alternative grouping. That is the purpose of the pretest, and each unit will have a pretest.

KATHY: If some people get better in the unit, can they move into the alternative group?

TERRY: I'm open to that. We can see how well you are working through the unit and how much explaining you might need.

Terry's commitment to practices that are equitably fair is ultimately revealed in a statement of regret: "There are some [students] that I've just felt like I could have done more for along the way".

Over the entire school year, I identified only three instances that I could not justify as either equal or equitable and, therefore, considered unfair. I present them, not to be critical, but to further illustrate how I interpret the principle of fairness in Terry's practice. The first relates to the weekly line leader. This is a coveted duty whereby students take turns leading the class to different rooms in the school and carrying the key-card that allows access to the school buildings. It is the only duty that Terry specifically assigned every Monday morning in the upper right corner of the whiteboard. Beginning impartially at the top of the class list, Terry progressed down the list in order, returning to the top when the last student had taken a turn. While this seems fair, the number of weeks in the school year was not divisible by the number of students in the class. Consequently, some students were assigned more opportunities than others. Holidays and professional development days created occasional short weeks, and when students were absent they missed all or part of their turn. Terry did not offset these inequalities, nor are they justifiable in relation to particular student needs.

The second example concerns in-class birthday celebrations. Initiated and organized by parents, they were simply accommodated by Terry during the school day. This resulted in notable differences. Some treats were large; others were small. Some were homemade; others bought. Some were accompanied by birthday napkins, candles, loot bags, siblings, and parents; others were unaccompanied and sent with the child in original packaging. Moreover, I am not certain if all of the children were able to celebrate their birthday at school, particularly those whose birthday occurred on a holiday or during the summer vacation. Again, Terry did not offset these inequalities, nor can they be justified in relation to particular student needs.

Lastly, a seating change that occurred within the first week of school might also be considered unfair. Terry established the first seating arrangement, giving priority to students' social comfort and well-being but with the following proviso:

> If they're good friends and they can work together, well then they can be with each other for the first couple of weeks of school. And if they don't work together well, then I separate them into different desk groups.

Seated together, Connor and Noah were rowdy and disruptive. Terry advised them: "I know it is difficult for such good friends to not chat when they are sitting together. But if this is too difficult for you two, one of you could choose to move to the other seat at this table. I won't make you. You can decide. But I don't want you to continue like this. It makes it hard for you both to learn". They remained together, but their behaviour did not change despite several additional reminders. The next day, Terry said, "Because you haven't learned, I am going to move Connor over here". With his poor eyesight, however, Connor could not see the whiteboard clearly from the new desk. Gabby volunteered to switch seats, and Connor returned to the front but at a different table from Noah. These moves are diagrammed in Figure 4.1.

Gabby was now seated at the back of the room, with two other female friends. Bonnie, whom Terry had intentionally seated with Gabby, was now at a table with three boys, none of whom were her friends. Although Bonnie did not complain, I wondered how fair this was for her and if Terry had been insensitive to Bonnie's need for the comfort of a friend during this first week of school. I asked Terry for her thoughts on how the situation had unfolded:

> I needed Connor to be here near the whiteboard because of his eyesight, and I didn't want him to be here because Noah was here. I asked one of students at this table, "Do any of you mind moving over there?" Gabby raised her hand and said, "I'm okay with it". And I said, "That's really nice of you". Instead of me, it being my decision, I wanted them to volunteer out of their own kindness, knowing the reason why he needed to be up there, and be willing to give up their seat. I wasn't sure that it would happen this nicely so early in the year.

Not wishing to risk my relationship with Terry or my position in the classroom so soon into fieldwork, I did not press the point that Gabby's act of kindness might have conflicted with sensitivity and compassion for Bonnie. As with the first two examples, this situation is not justifiable in relation to either equity

1. Conner's move away from Noah
2. Connor's switch with Gabby

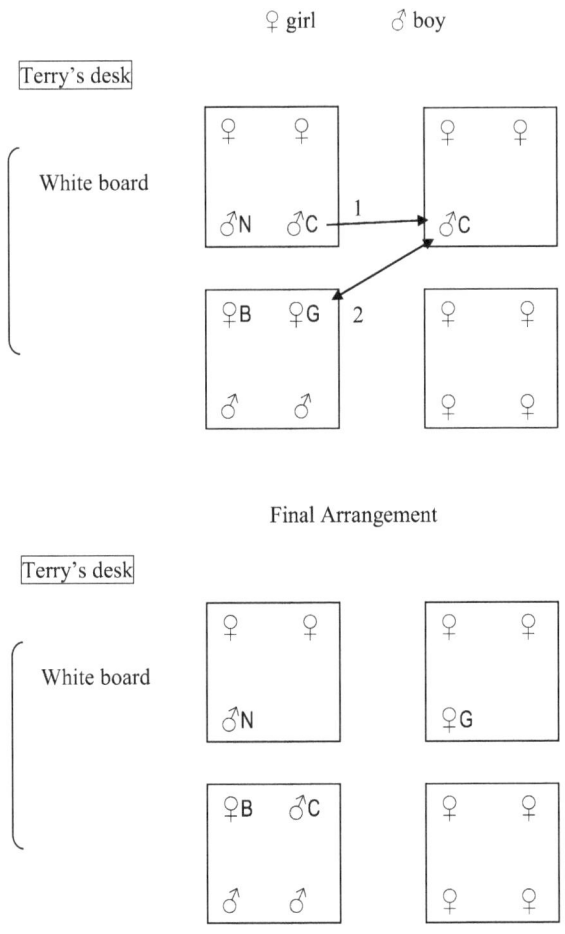

Figure 4.1 Seating change

or equality. Nonetheless, while likely unfair from a moral perspective, all three might be understood as logistically reasonable from a pedagogical perspective.

Respect

Respect was readily visible in Terry's conduct, behaviours, and practices, most notably in how she acknowledged everyone by name, expressed appreciation, engaged in presentations, trusted students, accommodated students' needs and moods, and avoided practices that might cause embarrassment. Regarding the first three—acknowledging everyone by name, expressing appreciation, and engaging in presentations—it is of particular significance, as a sign of respect, that Terry's manner did not differ between adults and children, the Head of School and support staff, or parents and colleagues. For example, to the Head of School she said, "Good morning Mr. Patrick. How are you today?"; to a visiting student, "Welcome Adam. What can I do for you?"; to a support teacher, "Hello Ms. Talbot. We're almost ready for you"; and to the school custodian, "Did you all say hi to Mr. George? He's here to fix Paige's desk". Acknowledging others by name extends beyond conventions of friendliness and politeness. It signals each as a person of importance and worth whom others should know. This was underscored when Terry mistakenly referred to a girl by the wrong name while substituting in a grade-three class. Terry corrected herself and publicly apologized, making sure to refer to the girl again, by name, before the session ended. Appreciation was also non-discriminately expressed. After library sessions, Terry said versions of "Thank you Ms. Laurie. That was a very interesting story". At the end of the first grade-two learning buddies session, Terry told the younger students, "It was lovely meeting all of you". In subsequent sessions, she said, "That was a wonderful meeting. Thank you". Lastly, during the peacemakers' training program, Terry had technical difficulties showing a video and eventually abandoned the activity, saying, "Thank you for your patience". These examples may seem trivial and insignificant. That is the point. Even in such small moments, Terry remembers to show appreciation. Finally, Terry fully and deferentially engaged in school and class presentations. During Monday morning assemblies, she rarely talked to the teachers with whom she sat and never brought her laptop, as others occasionally did. As with concerts, dramatic performances, and class presentations, Terry kept her eyes on presenters, responding with laughter, clapping, tears, or solemnity, as appropriate. Despite the more intimate setting of class presentations, Terry never interrupted. Her questions and comments at the end, however, demonstrated that she was actively listening and genuinely interested.

Respectful behaviours directed at the students, in particular, relate to trusting the students, accommodating their needs and moods, and avoiding practices that might cause embarrassment. Trust is an indicator of Terry's respect for her students, often manifesting with increased levels of student autonomy, empowerment, and freedoms. While doing seatwork, for example, students might disperse, at their own discretion, into the hallway, stairwell, computer lab, or library. Because direct supervision is limited, they are trusted to regulate their own behaviours and accomplish assigned tasks. Terry also trusts what students tell her, without feeling the need to fact check. When Frances was moving back and forth between the hallway and the classroom, Terry asked, "Why do you need to go back to your locker? Is it because you are cleaning out your desk?"

"Yes", Frances replied.

"Oh, okay then. Go ahead", she said. On another occasion, Terry asked Noah, "Were you here before the national anthem played?"

"Yes", he replied.

"Okay. Thanks". Noah was able to join the class without a late slip from the office.

Students took supplies without asking permission, including markers, geometry equipment, worksheets, coloured, lined or plain paper, art supplies, and the soccer ball. They were also allowed to borrow Sharpie markers, the stapler, rubber bands, thumbtacks, tape, and paperclips from Terry's desk drawer. If a student asked permission to do so, Terry replied, "You may. Help yourself, and thank you for asking". She never instructed the students to do so or reprimanded them if they did not. But she did request that supplies be returned so they are available for others. This was similarly the case regarding in-class library books, for which there was no sign-out process. Terry signalled trust regarding property in the first week of school when Connor asked, "Why don't you lock your cupboard, Ms. Kennedy?" The lock hung loosely from the cupboard door where Terry and I stored our purses and other personal belongings.

Terry replied, "Because I trust you". To my knowledge, that cupboard was never locked, although the classroom door always was when no one was there. Interestingly, Terry's trust in her students did not seem to extend to the school community.

There is a limit to Terry's trust in the students, however. For tests, they were asked to move the desks apart and scramble their orientations to face different directions. Terry explained, "So you are not tempted to talk or look at your neighbour's work". There were no complaints or perceptible expressions of surprise, even though Terry's language implied a lack of trust. This procedure was accepted as a school convention, rather than a moral concern, and a necessary reminder of test-taking expectations, given Terry's otherwise collaborative pedagogy.

Respect was also evident in how Terry accommodated a range of student needs and moods. On one occasion, for example, when Terry's class was preparing to write a test, Wendy's class was involved in an interactive activity that generated a fair amount of noise through the closed but adjoining door. Although the students did not complain, Terry was concerned this would compromise their ability to focus and be successful. I followed her next door, where she explained the situation to Wendy's students and requested their "consideration" and "cooperation" for the next 45 minutes. The exchange was not hostile, threatening, or punitive. Terry was kind and patient, demonstrating respect for both groups of students and the activities in which they were involved.

Further, as previously noted, the first seating arrangement sensitively and thoughtfully anticipates the students' social needs. Terry explains:

> I spoke with the grade-three teachers from last year to see who needed to be with whom, just for the first while. Or who didn't need to be near

whom. And then that's how I start off. In the beginning of the year, it's a new grade and a new teacher. In order to help make them feel comfortable, I like to make sure that they have somebody they can turn to or somebody that they're near, helping them feel comfortable. So they come into the class and can say, "Oh yeah, I'm next to you".

As the year progressed, expectations regarding class community became entrenched, and social considerations were no longer as relevant. Subsequent seating arrangements began to prioritize students' academic needs, such as those related to Noah's struggle with focusing, Bonnie's difficulties with independent work, and Connor's poor eyesight. Terry clarifies this, as follows:

They have to be able to work well, and some kids work better when they don't have to make an effort to change their position in their seats. For example, if you're facing the door, let's say Zeth and Bonnie, you have to turn around to look at some parts of the whiteboard. That extra effort may mean that when I teach at the whiteboard, I'm going to have to remind them to look up because they have trouble focusing. So those are the ones who have to face the whiteboard naturally.

Terry's commitment to accommodate these needs was no stronger than her commitment to accommodate social needs, demonstrating her respect for the whole person, not only the *student* part. This observation is also supported by how Terry accommodates the students' moods, sometimes forgoing her own agenda to do so. For example, when students appeared particularly restless, she spontaneously substituted a formal math lesson with an interactive, hands-on coat-of-arms activity. On another occasion, Connor declared, "I need to go outside. I need some air".

Several students concurred, shouting, "Me too". Terry agreed to an unprecedented 30-minute recess. Four girls did not wish to go out but were keen for a break, nonetheless. They remained in the classroom with me and played hangman on the whiteboard. Respect is noted in Terry's explanation for why she makes such accommodations:

We rush them. We rush them to get to the next level that sometimes we forget to just take a step back and breathe. And I do that, too. When it feels like they're not ready to understand something, and I get it from the way that they're looking and what they're saying to me and what they're not saying to me, I just say, "You know what? We'll just move right along", or, "We'll just talk about that another time. It's not important right now". And if it's me that has to give them the answer because they're not ready to do it or they're just not feeling in the mood maybe, fine. I'll just give it to them. You've seen me when I'm up there and I just say, "Never mind this for now".

Finally, respect was identified by Terry's desire to avoid practices that might embarrass her students. She is sensitive to this when soliciting participation during a lesson, for example:

> I kind of look at them and see whether or not they're paying attention before I even choose them. If they are paying attention and their eyes are on me, then I will call on them to answer. If they don't know the answer and look like they are going "hmm", I'll just say, "Can someone else help?" Something like that. There may be moments when the more sensitive kids probably think, I don't know the answer, and I'm a little bit embarrassed. But I'm hoping that they get used to it and just try. I'm hoping it's a safe enough environment that they'll just say whatever and help each other out.

Further, although Terry constantly reminded the students to tidy their uniforms after recess and before leaving for another classroom, she thoughtfully refrained from commenting when shirts were not ironed, had taken on a yellow hue, or were frayed. Tidiness is an expectation for students and within their control. Yet, the condition of the garments is the responsibility of parents and a reflection of home life. To note this, even privately, might be embarrassing for some students.

Lastly, Terry does not publicly admonish individual students, although she may briefly *call out for conduct of a particular kind* (Fenstermacher, 2001; Richardson & Fenstermacher, 2000, 2001). In a typical example, Alexander and Noah were acting silly while lining up for music class. Terry moved Alexander to the end of the line, away from Noah, asking him briefly, "Do you know why I moved you?" Although all eyes were on Alexander, the moment passed quickly, without further consequence. More in-depth rebukes with individuals or small groups of students took place privately, in the hallway or in the classroom when the class was elsewhere. This was also the case if the nature of the discussion was sensitive or personal. Sometimes I was asked to leave, as well. Terry's desire to protect the dignity of individual students was reinforced during a class wide rebuke, following an overly noisy session with a supply teacher. Several students assumed that Terry was referring to the behaviour of a particular student and turned in unison toward him. Noticing this, Terry said, "This is about all of you, not a single person. So your eyes should not be anywhere but on me". Semiprivate discussions took place in the presence of the class, at Terry's desk or a student desk. This usually occurred when a small group of students was involved in a situation that was already public or of relevance to everyone. The previously recounted anecdote involving Noah and Kathy illustrates one such incident. Terry's desire to avoid embarrassing the students resonates in her explanation for how she determines the venue:

> In some situations I don't do it privately, and in some I do. And where I draw the line I'm not always sure because I don't always do it

consciously. I may say to myself, "this is a good one-on-one talk" or "this is a class situation". How do I decide that? Well, if everybody or several of them need to hear about it because they will come across it themselves, then maybe it's worth it to work through as a class, without embarrassing anyone of course. But I'm hoping that they'll learn to feel less and less embarrassed about things that we talk about. I think they need to grow up and say, "If there are bad issues that we need to talk about, let's just talk about them openly, without feeling like it's going to make me feel embarrassed". Because we're all going to have these issues at some point.

Of the six methods by which Terry demonstrates respect—acknowledging others by name, expressing appreciation, engaging in presentations, trusting students, accommodating needs and moods, and avoiding practise that might be embarrassing—I find the last one to be the most deeply moral in nature, due to the many moral values it involves and the moral judgment it requires.

Kindness

Some expressions of kindness are formalized practices that Terry enacts year after year. For example, on the first day of school Terry placed on each desk a box of Smarties, a new pencil and eraser, and a note of welcome, saying, "I hope you had a great summer! I am very glad you are in my class this year. . . . I hope this will be a special year for you". To celebrate having completed the first test, Terry passed around a box of Girl Guide cookies, and for birthdays, Terry gave each student a designer pencil. The depth of Terry's kindness, however, was not revealed in these instances, but rather in the informal, spontaneous, and often personal nature of her exchanges with students. When Kathy did not bring a lunch to school, for example, Terry walked to the grocery store during her preparation period, explaining, "I don't want her going hungry, and I know that household is not terribly organized. It's okay. I'll just get her a soup". She was neither angry nor frustrated when Kathy's lunch was delivered later that morning and discretely put the soup in a cupboard. Similarly, Zeth did not bring spending money for the school's Global Village fundraising event. Terry gave him 10 dollars of her own so that his ability to participate would not be compromised. She showed no concern for being reimbursed. Terry would often walk to the end of the hallway to enter the drama studio through a second door so as not to disturb the group of peacemakers eating their lunch just inside the first door. During a library session, Terry noticed that Gabby and Mary were squinting from sunshine streaming through the bay window. Although they did not complain, Terry pulled the blinds, whispering, "To make you more comfortable". Similarly, Terry loosened Noah's tie, noticing it had been knotted too tightly around his neck. When Zeth had not yet returned from the December break, leaving Connor at a desk to himself, Terry said to him, "If you are feeling lonely over there, you can join another group". Such instances occurred continuously, too frequently, in fact, to enumerate. They were so deeply embedded in Terry's conduct that not even Terry was aware of her many acts of kindness.

The examples above are underpinned by moral values of sensitivity, thoughtfulness, and care. Three further examples seem to be more profoundly kind because of their additional association with empathy and compassion. The first relates to Mark, who had only recently arrived in Canada from Australia. Just before he was to present a reflective piece of writing at the school-wide assembly, Mark began to cry. Overcome with anxiety, he could not be comforted by Terry or his peers. In an unprecedented decision, Terry excused him from the presentation, foregoing her typical messages of courage and self-confidence. Toward the start of a math test, Terry noticed that Heather had tears in her eyes and was unable to focus. Heather had been involved in a confrontation at recess, and Terry believed this to be the cause. With a hand resting on Heather's head, Terry leaned in to coach her along. Gently but firmly, she said, "Panicking is not allowed. It's not productive. Come on. You can do it. Take one section of information at a time. What is first? Good. Now do the next part". Heather did complete the test, but Terry was not satisfied that it represented her best effort. The next day, she asked Heather to rewrite poorly answered questions. Heather was not excused from writing this test, although she was supported. The final example is recounted anecdotally.

> "I need to run something by you", Terry said to me during a quiet moment alone. A couple of weeks into the school year, Terry noticed that Bonnie was touching herself inappropriately in class, and not briefly, but for prolonged moments. "Have you noticed anything?" I had not. "Would you mind watching her a bit, and letting me know if I'm imagining this?" I agreed to do so, and by the end of the next day was able to confirm her observation. After a week with no improvement, Terry called Bonnie's parents, uncharacteristically asking me to leave the room while she did so. "I'm not looking forward to this call", she admitted with visible anxiety. I moved toward the door and closed it firmly behind me.
>
> Recess ended and the students returned in a flurry to the classroom, with me closely following. Terry seemed more herself as she launched into a social science lesson. It was not until after lunch that we found ourselves alone again. "So apparently Bonnie's mom was waiting for my call", she began. "She gets this same call every year about this time from the teachers". Terry briefed me on their conversation, relief in every breath. The family's paediatrician would be consulted and a referral to a specialist initiated. Bonnie's parents are confident, nonetheless,

that this behaviour is related to anxiety and not sexually motivated; after all, she is a *young* 10-year-old. Further, Bonnie did not appear to be aware of her actions. They were trying to gently and discretely raise her awareness and let her know this was private behaviour, meant for alone time. Bonnie's mother would provide her with a squeeze ball to keep her hands otherwise occupied during class. "She asked if I would just remind Bonnie to use it. So that's what we do", Terry concluded.

Over the next couple of weeks, Terry approached Bonnie's desk when she noticed her engaged in this way and placed the ball in her hands or matter-of-factly tapped her shoulder to redirect her attention. With the exception of the ball, there was little distinction between how Terry treated Bonnie and how she redirected the attention of others who were off task. And then, the behaviour stopped.

During this time, the other students seemed oblivious to what was taking place, and there was no perceptible change in Bonnie's demeanour or social situation. This is a credit to Terry's sensitive discretion. Terry also demonstrated courage by assuming responsibility for the situation. She could have deferred to Tom, since it was well within the scope of his practice as school psychologist. Alternatively, she might have chosen to ignore the behaviour or pretended not to have noticed. Given the eventual outcome, there would likely have been no harm in doing so. Yet, by her own admission, this would have been inconsistent with her personal values. One might argue that ritualized acts of kindness and kindnesses emerging from aspects of schooling and classroom life are an inevitable part of teaching practice. It should not be assumed that Terry felt a professional obligation to perform the more intimate acts of kindness. These are a manifestation of her character.

Honesty

With continuous professional and personal reflection, Terry becomes self-aware and self-knowledgeable, and avoids self-deception (Lickona, 2004). These are the foundations of self-honesty. Sean recalls, "What I've always admired is that Terry is self-reflective. She stops after a lesson and reflects, 'How did that go?' She's always the last to have her report cards done because she reflects, and reflects, and reflects". Terry concurs: "With every kid, with every year, there's always something that makes you question and reflect, and go, 'Okay. That's not working. At this point, what do I need to do?'" Reflecting on teaching practices serves as a gateway to more personal reflections:

I kind of had to learn things about myself, what I have to do. But that's an ongoing process of reflection. Have I done what I'm supposed to do too? You always look back. I always think about what I should have said.

More personal still, Terry reflects on the nature of her character in the context of her professional role:

> I'm not sure that I am always empathetic. It's maybe where I'm lacking. Because I think maybe if I were always empathetic, maybe I would have more patience. Or I maybe would not come down as hard on Zeth.

Similarly, Terry wondered if she should be more physically affectionate with the students, hug them more, for example, but decided that this would not be true to her nature, as Sean observes: "Terry's not gushy; she's not mushy, but the children know she loves them". This last statement, in particular, suggests that self-honesty facilitated Terry's ability to represent herself truthfully, or honestly, to others. She did so by publicly acknowledging shortcomings and weaknesses, allowing the students to know her on a personal level, and disclosing in-the-moment emotions.

Regarding public acknowledgment of shortcomings and weaknesses, Terry made the following confession to the class:

> Some people are really neat and organized. You go into their cupboards and all the things are really neat. I'm not like that. I need to see it, or I'll forget about it. So Ms. Bell helps me with this. She'll keep things for me so I don't forget about them.

Terry also admitted:

> I'm willing to let [the students] take over that because I know I'm not strong with my memory. I used to be able to know the entire schedule. I used to remember all of it. I can't do that anymore. So I have to rely on them to remember it for all of us.

This was regularly communicated to the students, with versions of "Who will remember this for us?" and "Do not forget to do this because I will forget to remind you". The students did not take advantage of such situations. Instead, they became adept at compensating for any lapses and preemptively reminding Terry of a variety of school and classroom events. Terry also acknowledged, "I am not always the expert in the classroom. I don't know everything". Accordingly, she sometimes asked, "Who can help me answer this one?" When introducing the spelling program to students, Terry announced, "We are using a new spelling program this year. While I get used to it, I'll have to refer to my notes. I'm sorry if that slows things down a bit for us". As a final example, Terry and Wendy both felt uncomfortable with the *pulleys and gears* building project in the science unit on structures. They invited a staff member to work with the students, saying, "Mr. Casey is going to help us out because we aren't very good at this stuff". In openly and honestly presenting herself as a flawed and developing person, Terry also demonstrated modesty and humility. She does not make herself the star of the classroom, or even a featured act. The students are the stars, and Terry allows and encourages them to shine, both in her presence and despite her presence, often beaming from the sidelines as they do.

Sharing selected aspects of her out-of-school life, Terry afforded the students more personal and intimate access to her. She posted several photographs on the bulletin board by her desk of herself with colleagues, family, and friends. Terry explained why:

> When they ask why [the photos] are there, I say, "Because they make me feel good, and I look at them if it's a bad day and it's a little bit of a relaxation point for me". But usually they don't notice them, unless they are up here asking me a question. During recess, when they're slightly more relaxed and they're not working, then they look around a little bit. It's nice for them to see that I do have a life. That it's not always about school, for me either.

The pictures are social in nature, rather than professional, but the poses themselves do not make obvious Terry's relationships with the other subjects. Previous classes were aware that Terry has a boyfriend, but I never heard a reference to this in the class I observed. Terry told me, with a chuckle, that some former students asked if they could be flower girls at her wedding. She refrained from such discussions, vaguely responding, "We'll see". While not dishonest, this answer demonstrates that Terry does have a limit to the truths she will reveal to students, something Fallona's (2000) research with teachers also identified. While allowing students to know her on a personal level, therefore, Terry is able to maintain a distinction between being friendly with students and being friends.

Finally, Terry is emotionally honest with students. This entails knowingly, and in some cases intentionally, expressing a range of emotions and allowing herself to be vulnerable in doing so. Terry explains, in the following three passages:

> Even in those moments when I feel like I am not at my best and I am getting frustrated more easily with [the students], I'm open. Even emotionally I'm honest with them. Yesterday, Connor wanted to ask the [Armed Forces] veteran how many "dudes" he killed. It bothered me a lot because at the Remembrance Day assembly there was a slide-show of 132 photos of soldiers that had been killed this year. Then there was a song that said they had moms and dads and children to say goodbye to. I was already feeling overly sensitive when Connor asked. Maybe I should have said, "I understand where you're coming from because I can see that it's a question you probably want to know the answer to". Instead I said, "Is that something that he is going to want

to brag about? Why would he want to do that?" I was emotionally out there. It was an honest response to how I felt.

So I said, "Here are your actions, and it's affecting me. I do actually go home and think about you. It's not like I let it go and I don't think about it. This is what you're doing. When I go home or when I wake up in the morning and have to go to work, it doesn't make me feel good". I said that to the class last year, more than once. Not to hurt them but just to make them aware.

I tell [the students] when I'm really frustrated, "You know what? I have about this much patience right now". So they know why I'm upset with them and why they cannot push my buttons right now. Because this is where I'm at.

Additionally, on the first day of school Terry publicly admitted to being nervous. Preparing for the Remembrance Day assembly, she grabbed a few tissues, declaring, "In case I cry". When the girls recounted stories of bullying, Terry empathized, "This makes me want to cry because I know how much this hurts". While privately disciplining Noah for kicking his grade-two learning buddy, she shed a few tears. Her eyes welled with tears of pride after Alexander read aloud to the class. Lastly, Terry reflects, "If something is humorous, I will laugh about it because that's the way it was and that's an honest reaction to it. I should give them an honest reaction". This is illustrated in the following anecdote:

It's mid-morning. Terry will return to the classroom a few moments after the students, having been outside on recess duty. The students know this and would typically assume their seats to read quietly, while waiting for her. I'm not sure whose idea this was, but as the students entered the classroom on that particular day, they were informed by others, "We're hiding on Ms. Kennedy. Hurry". With enthusiasm and much creativity, they tucked themselves into nooks and crannies, hoping to become invisible and desperately trying to remain quiet and still. Several crouched behind and under Terry's desk. Connor was behind the hallway door. Noah sucked in to squeeze behind my chair, between the file cabinet and computer desk. Frances went into a cupboard. No one sat at a desk. They waited. I watched from my perch, hiding Noah and trying myself not to laugh.

"Where is everyone?" Terry asked me when she returned, genuinely perplexed. I looked at her and shrugged with my eyebrows raised, eager to be convincing. Turning around she walked back into the hallway to look for her class. Unable to contain themselves any longer, giggles began to emerge. Hearing this, Terry came back into the room and laughed, as well. "Oh, I see. You are hiding on me. Very funny, you lot". Slowly, heads revealed themselves, and the students moved to their seats, thoroughly pleased for having pulled one over on their teacher.

Although Terry might limit the truths she reveals, she avoids deceit and telling lies. This was noted in her willingness to confess oversights and mistakes, give credit to others, and provide feedback to students. Oversights were often attributed to memory lapses. The following statements are typical of how Terry acknowledges them: "Yes. You need a pen for corrections. Sorry. I forgot"; "I forgot, didn't I? I'm glad you reminded me"; and "I'm sorry, Zeth. I forgot. We will work on it today, together". Terry also confessed to mistakes regarding academic work and took responsibility for corrective action. For example, she began a math lesson by apologizing for answering a question inaccurately the previous day. Before proceeding with the currently scheduled lesson, she corrected her mistake and made sure everyone understood. When the students did not do well on test questions related to perpendicular lines and congruence, Terry acknowledged that she had not taught that topic adequately. She excluded those questions from their score and retaught the concept. Relatedly, Terry declared, "Our field trip to the museum showed that Ms. Bell and I didn't do a good job teaching you about minerals. So we're going to do that now". When Mr. Roy publicly corrected Terry on a couple of points, while co-teaching a science unit with her and Wendy, she graciously and genuinely responded by saying, "Thank you Mr. Roy for clarifying that"; and "I didn't know that. Very interesting. Thanks for adding that Mr. Roy". When Terry did not provide enough coloured cue cards for all of the students to complete an assignment, she said, "This is my problem. I did not plan for enough. I'll have more cards by tomorrow". The final example involves a personal indiscretion, in which Terry admits to mishandling a situation and behaving badly, herself:

> There was a particular incident last year that I am not proud of. A boy was arguing with me and I argued back. Luckily it wasn't in front of everybody. Afterwards I apologized to him, and he apologized. He apologized first because I sent him to Jonathan. But then I told him why that didn't make me feel good—that whole exchange didn't make me feel good. And I cried in front of him and said, "I'm really sorry and I know that you're feeling this way too".

Reflecting on making mistakes, Terry says, "I'm human and I'm allowed to make mistakes. [The students] need to know that I can make mistakes. They don't have to see me as somebody who's perfect. I still require respect, regardless".

Giving credit to others, where and when it is due, also denotes honesty and truthfulness. Terry did so on many occasions, several of which were previously recounted as examples of showing respect through appreciation. In addition to those, Terry announced to the class, "Ms. Laurie did a wonderful job putting this reading list together for us. It takes a lot of time and effort to read all of these books and put down a little description for you". When another teacher managed a behavioural problem between their two classes, Terry said, "Carol did a fantastic job of talking through this with the students". Credit was similarly given to students. The following statements represent a sampling of many such instances that occurred throughout the school year: "Yes. Put your chairs up. Good thinking, Connor"; "Connor has a really neat trick for multiplying numbers in the 90s that his nanny taught him. I wanted him to show you, himself"; "Look how Gabby has done it. That is another good way to set things up. Thank you, Gabby, for sharing that with us"; "Kay brought up something that was quite perceptive. Please be as detailed as possible. As Kay pointed out, there's more than one time this happened"; "Sammie has a really good idea as another option"; and "Thank you for asking those questions, Zeth. That's helpful for everyone". Once again, there was no discernable distinction between Terry's willingness to credit adults and her willingness to credit children. Credit was based on merit, honestly, fairly, and respectfully given to whomever was deserving.

Lastly, Terry prioritized honesty and truth in providing students with both positive and negative feedback. She explains why:

> The honest reactions, the responses that I have to a piece of writing, for example, are true. It's my real response. If I fake it, I think somehow they will get that it's not a real response. They see what I look like when I really mean something. If I don't mean it, I'm pretty sure they think, "She didn't really mean that".

Regarding positive feedback, in particular, Terry said, "I don't like compliments that aren't real. When I tell them that they've done a great job, I really do mean that they've done a great job". This is also the case with negative feedback. Terry recounts the following incident by way of illustration:

> The other day, I had to tell them straight-up, "As a whole class and around Christmas time, you're supposed to be giving. You're supposed to be more kind. You're supposed to be more compassionate. And yet, you are ganging up on each other and complaining about each other. And it's just not nice. It doesn't make you a nice person". And I will tell them, "You're not a nice person right now". Those kinds of things they may not be used to hearing from a teacher because teachers try to be encouraging all the time. But if I see that they are being nasty or brutal to someone else, I'm going to tell them they are not being nice.

Rarely, however, was the feedback purely negative. "If [the work's] not good enough then we always pick on the good stuff and say, 'Here's how you can add this and this and this'", Terry explained. Positive comments were similarly intentional while giving negative feedback on behaviour and social-moral interaction:

> I may say something like, "Zeth, I really like who you are, and I think it would be a real shame if you continued this way because you'll end up alienating yourself, and you won't have friends. And that would make me feel sad for you".
>
> I wanted to tell [Heather], "You are happy. You are strong, and you can deal with things. So deal with them. When you talk to your mom, be honest". Because Heather tends to go home and tell a different story. If her mom saw her on a day-to-day basis at school, then she wouldn't be so frantic about things that she thinks are going wrong in Heather's life.
>
> I told [Sammie], "You're a wonderful person, Sammie. Be happy with who you are. Others will respect you more, though, if you just let go a little and tell it like it is. Don't be so concerned about denying everything. It will be okay".

This balanced honesty engenders trust, as Terry notes:

> Sammie never quite trusts the good things others say about her. So she comes to me. For example, in the personal comment sheet they all had to write one compliment about each other. She was reading the top one, and it was in cursive and very messy. She wasn't sure if it was a compliment or an insult. So she came and saw me and asked, "Is this a compliment or an insult?" And I said, "That says you have a strong voice. That means a powerful singing voice. It's a compliment". She trusted that, but only because I told her.

Terry reflects more generally: "One of the comments that came back to me from a former student is, 'That's why I like Ms. Kennedy. She just tells it like it is'. It's true".

A Note on Representation

Organizing this discussion around fairness, respect, kindness, and honesty should not imply a hierarchy of moral values regarding Terry's conduct, behaviours, and practices. There is no empirical evidence to support these four as particularly prioritized over other moral values, just as there is no support for Terry's claim of prioritizing integrity:

> Integrity is the one that's the most difficult and the one that's most important. It is doing what is right. And there are so many grey areas in different situations as to what to do and what the right thing would be.

That's the challenge and that's the difficulty of integrity. But that's what makes it so important. Because in the end, what's the glue that holds that person together, in order to be able to pick the right thing to do?

Integrity was the only school value, in fact, that Terry did not address in a formal lesson, and it was rarely the subject of a direct message to students.

Further, it should not be assumed that moral values absent from this discussion are irrelevant to Terry's conduct, behaviours, and practices. Rather, they are often inherent in or closely associated with those that have been identified, sometimes inextricably so. For example, integrity is deeply embedded in the discussion above by its association with consideration for others, good judgement, self-knowledge, and honesty (Lickona, 2004; Ryan & Bohlin, 1999). Terry identifies this when reprimanding students:

You will all be admired and respected more by people if you are able to say what you did. That is being a person of integrity. Explain to me that you made a mistake. That is much better than me having to get it out of you. I have to be able to trust you to behave with integrity by this point in the year. I cannot be there every time.

In addition, empathy underpins expressions of fairness, respect, and kindness and is at the core of the Golden Rule. Accordingly, Terry often asks her students versions of "How would you feel having it done to you?" Yet, she tends to understate empathy while reflecting on her practice, saying, "I don't relate to empathy or connect to it as I do with integrity".

Therefore, in describing Terry as a moral person I have used my discretion and best judgment to classify her conduct, behaviours, and practices according to certain moral values and to identify connections and associations with and among several others. The latter is inevitably underrepresented, however. Another investigator might reasonably attribute different moral values to the examples or use a different structure for representation. It is unlikely, however, for anyone to disagree that Terry typically expresses the range of moral values identified above.

Expressing Moral Values as Moral Instruction

When teachers express moral values in their conduct, behaviours, and practices, they integrally and in some cases intentionally demonstrate for students how one *ought to* behave (Campbell, 2003; Hansen, 1993; Jackson, Boostrom, & Hansen, 1993). This is referred to as modelling. Modelling moral values assumes a direct connection between the moral character of teachers and the moral development of their students (Fenstermacher, Osguthorpe, & Sanger, 2009; Osguthorpe, 2008). As such, it blurs the distinction between teaching morally, the focus of this chapter, and teaching morality, the focus of the next. Fenstermacher, Osguthorpe, and Sanger (2009) observe, "The teacher is engaged in both teaching morally and teaching morality, as she is acting in a morally upright manner and making her manner an object of instruction" (p. 9). Accordingly, modelling is widely promoted as a moral education strategy in both character education (Lickona, 1991, 2004; Ryan & Bohlin, 1999) and care ethics (Noddings, 2002, 2008, 2010), and has been identified as such by The Moral Life of Schools Project (Jackson, Boostrom, & Hansen, 1993), The Manner in Teaching Project (Fenstermacher, 2001; Richardson & Fenstermacher, 2000, 2001), and The Moral and Ethical Bases of Teachers' Interactions with Students study (Campbell, 2003, 2004; Campbell & Thiessen, 2000, 2001).

In the early years of teaching, Terry modelled morality for students simply by being a good and righteous person. Modelling, therefore, was generally coincidental, as she focused instead on becoming adept with the mechanics and technicalities of her practice. As Terry "got more comfortable teaching", she became intentional in making morality visible to students as moral instruction. This coincided with her growing interest in moral education. Research for the Character Development Program helped to characterize her present position, which she articulates as follows: "Everything that the adults do, the children see. And regardless of whether or not you see them watching you, or hearing you, they are. So you might as well show them what you want them to see". Terry illustrates this with the example of demonstrating patience when making mistakes:

> I make mistakes and it's okay to make mistakes, and I want them to know that it's okay for them to make mistakes if it's okay for me to make mistakes. I'm not going to feel badly if I make math mistakes, so they shouldn't feel badly if they make math mistakes. I know some of them are still very concerned about making mistakes. Some of them will always be harder to get out of that mode. Kay and Pia don't like to make mistakes. I'm hoping that if I make mistakes and I don't make a big deal out of it, then they might become more patient making mistakes themselves.

A similar outlook applies to how Terry responds to interruptions during the school day, including broadcasted announcements, staff and student visitors, and phone calls. These occurred randomly, during lessons, tests, or seatwork, and when the students were reading aloud or giving presentations. Terry says:

> If I react in a negative way to whatever's happening here or there in terms of external interruptions then it gives them permission to do the exact same thing. Then a phone call comes and it may be an important phone call, and all they have left is just a negative reaction to it because they see me do it. And it could be somebody who needs my attention. So I can't just shrug it off. I know they're watching all the time, and any time I look angry or unhappy about something they pick it up, and then I don't want them to treat it in a negative way.

Terry was particularly tested on this when her co-teacher, Wendy, was unexpectedly absent for three days. There had been no time to prepare the supply teacher for the many homeroom and teaching responsibilities. Consequently, she required considerable support from Terry, often spontaneously appeared in the adjoining doorway, saying, "Sorry to interrupt, Ms. Kennedy, but I have another question". Each time, Terry paused her own activity to provide assistance, saying, "That's okay". I never observed negative behaviours from Terry, such as rolling of eyes, complaining, sighing, or otherwise expressing frustration or impatience. Some might argue that in prioritizing interruptions and allowing her attention to be diverted from the students, Terry was, in fact, disrespecting them and the activities in which they were engaged. She acknowledged and mediated this possibility, saying upfront, "Excuse me for a moment". After, she apologized, saying, "Sorry about that. Please continue".

Modelling also lends support, credibility, and authenticity to moral messages and lessons imparted by other strategies (Campbell, 2003; Fenstermacher, 2001; Fenstermacher, Osguthorpe, & Sanger, 2009; Richardson & Fenstermacher, 2000, 2001), including those described in the following chapter. Terry recognizes this: "If I teach morality without being a model of a moral person, students won't take me seriously. If they are reflecting on the ideas, they will see that I'm not following them, so why should they". By way of illustration, Terry notes her participation in community service: "I don't want to be a hypocrite, which is why it's important for me to find ways to help the community. I teach them to contribute, so I should, too". Similarly, Terry upholds the same behavioural expectations she communicates to the students, including, "When someone says good morning, you say good morning back. That's being respectful"; "Mistakes are natural. Don't get frustrated. Just do it again"; "Mistakes are okay. If you didn't make mistakes you would have nothing to learn. I learn from my mistakes". The significance Terry places on modelling morality is underscored by her remorseful recollection of the following incident:

> When I lost my temper several times last year it always felt bad. It always made me feel badly thinking about it because I'm the role model, and I'm the adult. And if a kid is losing his temper in front of me and arguing with me, then, as a role model, I shouldn't be arguing back. But I did.

Terry was aware that her behaviour undermined the messages of respect, compassion, and patience she regularly imparts to students and that it demonstrated values antithetical to moral conduct and interaction.

NURTURING CLASS COMMUNITY

Learning environments anchored in sentiments reflecting human flourishing, dignity, welfare, and well-being are often portrayed by the metaphor of community. Accordingly, Solomon, Watson, Battistich, Schaps, and Delucchi (1996) characterize community as a social grouping, where members experience a sense of connection, commitment, belonging, caring, and support, and where members contribute to and participate in collaborative decision-making, planning, and deliberation toward shared purposes and common goals. This definition encompasses Nucci's (2009) description of a moral classroom climate, which fulfills four basic needs of children—belonging, autonomy, competence, and fairness. It also infers the three communal processes Furman (2004) proposes in promoting an ethic of community—knowing, understanding, and valuing others in the community as unique individuals; full participation and inquiry by everyone in the community; and working toward a common good. Taken together, these characterizations suggest that community might be understood as both product and process.

Terry's efforts to nurture a classroom community, in fulfillment of her vision, can be similarly understood. As a product, Terry fostered relationships among the students. Relationships support the community by helping students know, understand, and value one another, and experience connection, commitment, belonging, caring, and support (Johnston, 2006). As process, Terry facilitated student collaboration. Collaboration supports the community by enabling students to autonomously and competently contribute to classroom life; participate in decision-making, planning, and deliberation; and pursue shared goals (Furman, 2004; Oser, Althof, & Higgins-D'Alessandro, 2008). Terry's practices that foster relationships and facilitate collaboration are detailed in the sections that follow.

Fostering Relationships

Fostering relationships among students was an immediate priority for Terry: "If I don't build relationships right from the start, then later on it's going to be really hard to backtrack and try to build on that". Accordingly, she initiated two *getting to know you* activities on the first day of school. The first, a class-wide conversation, began with Terry sharing a small fact about every student, including, for example: "I know something about Connor. I know that Connor loves hockey"; "I know Heather has a sister"; "I taught Bonnie's brother. How is he doing, Bonnie?"; "I know that Kathy has pets"; and "I know that Sammie spent the summer in Europe". The students fully engaged, laughing and offering their own comments. Gabby added that her sister and Heather's sister were friends, and Kathy listed all of her pets and their names. The second activity revolved around a questionnaire. The students were asked to survey as many classmates as possible, on siblings, pets, hobbies, talents, extracurricular activities, languages spoken or understood, and food preferences, among other subjects. After several minutes, the students returned to their seats, and Terry initiated another lively but shorter discussion, asking, for example, "How many of us like pizza?" and "What kinds of talents do we have?"

Both activities encouraged the similarities and differences among students to emerge. Recognizing diversity among students' skills, talents, and experiences was particularly important as Terry increasingly relied on the students to offer, request, and accept peer support on a variety of academic and social issues for which they were differentially inclined. Pia was acknowledged as the class artist, for example, and was called upon to advise or assist others with their drawings. Connor is accomplished at math. He clarified questions for his classmates and sometimes corrected their worksheets. Kathy is socially aware and good at conflict resolution. Applying the peacemaker protocol with skill and sensitivity, she helped the girls, in particular, work through interpersonal problems. Terry ensured that every child was validated for his or her exceptionalities and had a role to play in classroom life. While pedagogically effective because students are not entirely dependent on Terry for support, this expectation is also moral in nature as it promotes the values of a class community. Explicitly, this includes helpfulness and assuming responsibility for others. Implicitly, it entails inclusiveness, acceptance, and care.

Terry's efforts to foster relationships continued throughout the school year, facilitated by seating arrangements and cooperative activities, both of which brought different groups of students together. Of nine seating arrangements, approximately one per month, only the first intentionally placed friends together. Beginning with the second arrangement, students were expected to intermingle with whomever they were seated. In the third arrangement, for example, Noah was placed at a table with all girls. Alexander, Connor, Mark, and Zeth also sat at otherwise *girls'* tables in subsequent arrangements. Rarely did the boys associate with the girls at recess, generally sticking to themselves and playing soccer with Wendy's boys. Yet, at these tables the boys and girls interacted positively and willingly. Particularly revealing, the fourth seating arrangement placed Zeth and Sammie at the same table, side-by-side, despite their having been in open conflict and publicly declaring their dislike of each other. Within the hour, they were chatting and exchanging friendly comments. Terry privately remarked: "I had hoped it would come to this point. This shows that they are starting to see themselves as a community. They are coming along".

The orientation of desks was also a consideration for fostering student relationships. The small classroom size limited the possibilities to two orientations, diagrammed in Chapter Three in Figure 3.1. For most of the school year, four desks were placed together, creating small working groups among the students. The desks were rearranged on three occasions into a horseshoe pattern opening toward the whiteboard. Students could not work in groups with this arrangement, only with the person on either side of them. Although the horseshoe challenged Terry's collaborative pedagogy, she appreciated another social benefit: "I can see everyone's face, and everyone can see everyone else".

Ending the year in this orientation was intentionally symbolic of how they had become a community, a single unit, rather than smaller groupings.

The class was also divided into groups and partnerships for cooperative activities. Occasionally, groups were organized according to the school's house system. Students were assigned to particular houses to maintain a balance across the campus and across grades. But the distribution in each class varied. In Terry's class, the houses were unevenly represented and generated uneven groups. More often, therefore, Terry designated the groups. Sometimes her criteria related to academic goals and the groups' ability to successfully complete the task. At other times, she expressed criteria related to pro-social and moral goals: "I like putting them in mixed abilities groups because they have to learn to be with a variety of people and abilities. It may teach them to be more helpful or to be more patient, respectful, and responsible." Partners were often randomly assigned using playing cards and numbered pick-me sticks. Students were asked to pick a card from different decks or a stick from different sets and find their match. Alternatively, Terry might pick two sticks from a single set and match their numbers to the students' numbered position on the class list. Regardless, she coached the students to respond respectfully and sensitively when partnerships were assigned:

> Keep your feelings to yourself about who your partner will be. Remember that your body language can give away your feelings, too. Don't slump. You wouldn't like it if someone said "oh, no!" about you or slumped in their chair with a frown. You never know what a great partnership you can make. It also isn't nice to shout, "Yeah!" That could make other people feel bad, too. Remember, good friends may not make the best project together. This is an opportunity. Your best friend may not be the best person for you to work with.

Despite preemptive efforts to avoid hurt feelings and foster functional groups and partnerships, Terry closely monitored the groupings and occasionally made changes, declaring, for example, "This is not a good idea. Let's pick again". She explains:

> Sometimes you'll notice if I'm doing pick-me sticks during the year and I come up with a pair that isn't working, that I know isn't going to work out right, I say to them, "At this point, I'm going to make a different choice because at this point in the year you're not ready to work together".

Working together involves cooperation and respect: "I know how difficult it would be if the partnership did not work out well. So I knew I had to put people together who would be able to cooperate respectfully, more or less". As the year progressed, Terry tended to leave a group or partnership intact, with a reminder or warning, despite reservations. In the following example, the warning was based on behaviour related to completing the task and was not directly moral in nature, although moral values were implied:

> Sometimes I give it a chance. If it's Noah and Connor with the pick-me-sticks, for example, I give them a heads up. And I say, "You two get to be pairs. However, here's what I'm looking at. If you two are working together well, it means that this and this get done, and done well. That means I don't have to come up to you several times during my walk-abouts and say, 'Focus, you're chatting too much'. If I don't have to do that often, then you get to stay together and work together. But if I have to continue telling you and reminding you to focus, then I will move you apart".

These opportunities for students to variably interact with their classmates do not inevitably foster relationships among them or nurture a class community. Maximizing the positive potential of seating arrangements, desk orientations, group work, and partnering and minimizing the negative potentials that Oser (1994) identifies with group work, in particular, also involved the consistent imparting of pro-social and moral values. When Terry announced, for example, "Don't rush. We are all friends here" and "This is not a competition among you", pro-social values of camaraderie and teamwork were conveyed. These values are underpinned by moral values of inclusiveness and helpfulness. When Terry asked for a show of hands in response to questions regarding the recent math test, including, "How many of you got them all right?"; "Who got one or two wrong?"; and "Who got more than that wrong?", she encouraged values of openness and mutual support, along with moral values of honesty, trust, and respect. The students did raise their hands, accordingly, even offering reflections and comments. On one occasion, Connor called out, "I got it wrong, and I know exactly where". Weaker math students similarly participated. Bonnie, for example, raised her hand to indicate when and where she had not done well and would publicly ask Terry how to interpret results in the lower scoring range. This suggests that Terry's values messages had been internalized by the students and were operational in the classroom, creating a psychologically and socially safe environment. Further, there was no evidence of classroom-based cliques, despite close out-of-school friendships between Mary and Gabby, Sammie and Pia, and Noah and Connor. Although not overtly discouraged, these relationships were simply not relevant in the context of Terry's pedagogy and classroom life.

Facilitating Collaboration

In a context of community, Terry equates collaboration with the sharing of ideas:

> If it's a community, then we all have to have some say, and everybody has to be able to have their ideas out there. It's just like working in a

group or in pairs—if you want to come together in any kind of work situation or social situation you need to share ideas and thoughts about it.

To facilitate sharing, Terry empowered her students, explaining, "If I never gave up my control, I feel that they would never have the confidence to be themselves". Consequently, students were empowered in many areas of classroom life, including seating arrangements and aspects of the classroom's physical conditions and appearance. Most of the seating changes were initiated by the students, with comments such as, "Can we change seats?" and "Time to change tables". Terry usually agreed, and within a week of the request a new arrangement was collaboratively determined. Paige, Frances, and Bonnie consulted with Terry on the second arrangement. They were not specifically selected for this task but had initiated the change and took ownership of it. Subsequent changes included the participation of more students or the entire class. Regardless, Terry guided the process with variations of the following values messages: "Friends can sit together, but you have to make sure you are making a good choice about your ability to work well together, too"; "You know what your needs are, so consider them"; and "Be mindful of different students' needs". Students were encouraged, therefore, to simultaneously advocate for themselves and accommodate each other. Advocating for themselves promoted moral values of self-confidence, self-respect, courage, integrity, and honesty. Accommodating others promoted moral values of acceptance, kindness, thoughtfulness, fairness, respect, and sensitivity. Terry revealed this moral intention when she said, "Instead of me, it being my decision where they sit, I wanted them to volunteer out of their own kindness". As with group work, however, Terry kept a close watch and occasionally intervened. The following anecdote illustrates how one seating arrangement was negotiated:

"Time for a change!" Sammie declared. Terry nodded in agreement. She asked Paige, Noah, Mary, and Gabby to head to the whiteboard with erasable markers. Mary drew a diagram of the four desk groupings, while the other three stood by expectantly. The rest of the class settled into chairs. "Consider who needs to be near the front because of seeing or focus issues", Terry reminded the students from Paige's desk, where she had seated herself.

Sammie spoke up first, "I'm not good with projects or spelling. So I should be near the front".

Terry replied, "Remember that you can get help for that by coming up to my desk, too. That's really not the same thing as having to sit near the front all the time".

"I need to be up there because I can't see", Connor called out.

Frances concurred, "I also have trouble seeing. I can see where I am now, so I should stay here". The four students at the board placed initials on the diagram of desks, accordingly.

"Bonnie needs to sit near my desk", Terry matter-of-factly added. I turned to Bonnie, searching her face and body language for a reaction to this potentially embarrassing declaration and wondered if it was a breach of trust on Terry's part. Bonnie showed no discomfort. Perhaps she hadn't heard. Perhaps she was okay with everyone knowing this. Perhaps they already knew from previous discussions. Regardless, the comment was passed over quickly with a barrage of other student needs. "Remember also", Terry interjected again, "those easily distracted shouldn't face the door. I think about this when I do seating plans. You need to think about this too". Terry then sat back and watched. Initials were written and erased as information was randomly called out. The students at the board tried to accommodate all requests while also considering their own.

After several moments of lively discussion, Terry suddenly sat up and announced, "There is a private concern, and I'm going to say that Sammie shouldn't go where you have put her". Kay volunteered to switch. Gabby erased Sammie's initials, replacing them with Kay's. No one asked why. They simply moved on. My mind was racing, however. What could the nature of this private concern be that it was more private than Bonnie's need? Sammie gave nothing away. She just looked blankly at the board. Terry's voice interrupted these thoughts, as the process continued. "I haven't heard from the boys, except Connor. Mark, where are you going? Please go up to the board and put your initials down". Predictably, the boys sat themselves together, prompting Terry to warn them, "You boys can't be silly though". Writing, consulting, erasing, and rewriting continued with controlled chaos until all positions had been successfully negotiated. Then the students rotated their present desks, belongings and all, into the new groupings.

Not all of the seating arrangements were as efficiently and positively collaborated. While discussing the eighth arrangement, at the end of April, Connor and Noah commented to each other that they did not wish to sit with Zeth. It was loud enough for most of the students to hear, including Zeth. Terry reacted swiftly and assertively:

Stop right now. Although Zeth is smiling, his feelings are hurt. Think about what you have said and how you would feel if that was said about you. I'm stopping this. It isn't going well anymore. I don't like this at all.

Terry facilitated a do-over the next day, rather than determining the seating arrangement herself. This reinforced Terry's expectations for kindness, respect, and sensitivity, her confidence that the students could meet these expectations, which they did, and her commitment to this collaborative process. The ninth and final arrangement was determined independently by the students, having put the previous difficulties in the past.

Students were also empowered to control some aspects of the class-room's physical conditions at their own discretion and initiative. Terry signalled this by announcing, for example, "If the sun is shining in too brightly for you, feel free to close the blind". Accordingly, when the students found it noisy in the hallway or adjoining classroom, they closed the doors. When the room became stuffy, they opened the doors or windows. When they were hot, they turned on the ceiling fan or turned off the lights. When they were cold, they closed the windows or shut off the fan. The students did not seek Terry's permission for making these changes but were encouraged to consider the comfort of their classmates, as well as their own.

Students also influenced the classroom's appearance. The year began with clear counter surfaces; the clothesline dotted only with red, green, yellow, and blue pegs; and bare bulletin boards, except for the themed borders around their perimeter. Several brightly coloured posters were strategically affixed to the limited wall spaces. Learning materials were thoughtfully situated throughout the room to create stations for specific resources and supplies. Books were neatly aligned with spines facing out. In the early weeks of school, this physical space strongly reflected the presence of its teacher. As the year progressed, however, the students' presence emerged and threatened to eclipse the teacher's. Except during lessons, student handwriting dominated the whiteboard. Counter surfaces were strewn with creative projects, structures at various stages of construction, and oversized written work that did not fit neatly into cubbies or folders. Completed assignments were strung along the clothesline, back to front, and stapled to the two bulletin boards at the back of classroom. The windowsill exhibited forgotten pencil cases and other mis-placed personal items. Colourful and themed water bottles adorned students' desks. Even Terry's personal space was not exempt from the students' presence. On the bulletin board by her desk, which was designated for personal and professional materials, Terry also posted the class photograph, as well as drawings, letters, cards, and poems that students had given her. This sym-bolically connects Terry with the students and designates the entire classroom as shared space. The overall effect was reminiscent of a child's bedroom, ini-tially conceived and staged by parents but gradually evolving into the orga-nized chaos that reflects the child's ownership and energy.

Community Achieved

Two events at the end of the school year symbolized the class community that had formed. In a mock ceremony, Terry presented each student with a small gift, such as a lanyard, bookmark, magnet, marker, or pencil, as well as a personalized *Sillies Award*, as follows:

Pia, shiny shoes;
Frances, losing stuff;

> Zeth, biggest hair;
> Mary, best pigtails;
> Heather, cuts and bruises;
> Alexander, coming out of a shell;
> Kathy, love of perfection;
> Connor, loudest voice;
> Bonnie, losing pens and pencils;
> Noah, leaving the water bottle everywhere;
> Gabby, quiet as a mouse;
> Kay, community service;
> Paige, willing to sit anywhere;
> Mark, best accent;
> Sammie, longest stories.

The students laughed and nodded their heads in agreement, as each shook Terry's hand and received the gift and certificate. There were no snide comments or snickering and no evidence of discomfort or embarrassment. The students seemed to understand and accept the activity's goodwill and good intentions, likely because the classroom culture was established on caring relationships and moral values of respect, kindness, trust, and acceptance. In this context, poking a little fun at oneself and others is simply that. The second event focused on the class, rather than individuals. Terry and the students watched a slideshow of photographs that Terry had taken throughout the school year. Every child was represented several times in a variety of in-class and out-of-class activities and in various peer groupings. Many shared experiences were recollected with commentary, laughter, and a few tears. The students received their own copy of this slideshow as a memento of their year together. Thus, in the pursuit of common goals, shared experiences, and collaborative processes Terry did not lose sight of the individual child. She encouraged the unique qualities and abilities of each to flourish and be expressed in relationship with others and by participation in classroom life. Achieving community for Terry, therefore, involves navigating a balance between the individual and the collective.

Community as Moral Instruction

Nurturing a class community blurs the boundary between teaching morally and teaching morality by providing a context in which morality is inevitably promoted and necessarily expressed (Lickona, 1991, 2004; Noddings, 2002, 2008; Nucci, 2009; Richardson & Fenstermacher, 2000, 2001; Ryan & Bohlin, 1999; Wynne & Ryan, 1997). Care, kindness, trust, honesty, helpfulness, responsibility, sensitivity, respect, acceptance, and inclusiveness, among other moral values, are involved in establishing and sustaining the relationships and collaborative processes that define Terry's class community. Terry seized many opportunities for imparting such moral messages. Helpfulness and responsibility were particularly emphasized, in support of Terry's expectation for students to recognize a need, operational or individual in nature, and take responsibility by offering to help. Operational needs are daily tasks that support classroom life and activities, such as handing out and collecting books, worksheets, folders, journals, and supplies; organizing garbage and recycling materials; wiping desks and sweeping the floor following lunch; tidying the computer lab, drama studio, music room, gym, library, hallway, and other shared spaces before relocating; running errands; answering the classroom phone when Terry is unavailable; cleaning the whiteboard; and clearing counter surfaces prior to the weekend. Individual needs are relevant to particular students. Related to academic activities, help might entail coaching prior to a presentation; performing as an extra in another student's book talk skit; explaining or clarifying questions; and correcting and editing each other's work. Individual needs might also be more personal in nature. For example, a student might want company while on an errand to the office, another classroom, or elsewhere in the school; assistance while tidying a locker or cubby; or an extra pair of hands if incapacitated.

These opportunities for helpfulness are not formalized by a duty wheel or other system of assigning and rotating responsibilities. Terry rationalizes her decision as follows:

> Not having assigned jobs does seem to reinforce in a more natural way that we are responsible for helping each other, whenever. Because if they do have labels, if they do have jobs, what they tend to say also is that, "It is not my job to do that. It's so-and-so's job". Then I need to say, "But, can you not help out?" No one can say, "That isn't my job", because it is everyone's job to help. Something like that would undermine the kind of community that I'm trying to build because I think that it's very narrow.

Terry does not want the students to empty the recycling bin, for example, simply because it is their turn to do so or to assist a classmate out of obedience. This would undermine students' moral motivation by removing the necessity for sensitivity, kindness, thoughtfulness, empathy, and compassion, moral values that underpin Terry's interpretation of helpfulness and responsibility in a class community. She notes:

> So that's basically compassion for others because without the compassion for others, and if you are only interested in what is happening to you, you can't be aware and observe what people need. So being helpful to others means taking them and yourself a little bit out of who you are and saying, "Wait a minute. Not about me. What do I need to do?"

Instead, Terry directly communicated her expectations with a general reminder to "help each other along" and with in-the-moment requests and suggestions. Requests included, for example: "I would like two volunteers to hand out math books"; "Since you are up, would you mind getting another one of these from the library please?"; "Would you mind booking the laptops for two o'clock today?"; "Would you mind closing the door for me please?"; and "Can you please go next door and ask Ms. Bell if we can borrow the large protractor?" When Heather broke her foot and was hobbling on crutches for two weeks, Terry relentlessly asked the class, "Are we looking out for Heather and helping her with her foot?" and "Who is helping Heather right now?" As a final example, one Friday afternoon Terry suggested the following:

> If your desk is too messy, you can help each other or ask someone for help. Some people are just more tidy than others, and that's okay. If that is you, then offer help to others. Ask for help and offer help.

By considering herself a member of the class community, Terry also modelled behaviours associated with these expectations. On one occasion, she instructed the students to "have a look on the floor. Get down on your knees, *like me*. Pick up anything you see that is for garbage or recycling". On another occasion, Terry tidied hats, mittens, and boots that had become scattered throughout the hallway. "I have to let them know that I will take responsibility, as well", she acknowledged. Further, as a member of the larger school community Terry is helpful to her colleagues. She always accompanied her students across the hallway to their library sessions with Ms. Laurie and remained with them the entire time. Terry explained:

> I don't think it is right to just leave my class there. This is my class, and I'm responsible for them. I should be here to help Ms. Laurie. Some teachers use [this time as preparation] and they don't come back. Ms. Laurie needs to be able to help the students find books. I sign them out for the students so she can do that.

This was similarly the case for concert rehearsals and computer sessions, during neither of which Terry had any teaching or class management responsibilities.

Most of the students assumed a range of responsibilities throughout the year and learned to spontaneously help each other. Connor was often heard asking, particularly during math, "Do you need help? I can help you". Despite hobbling on crutches, Heather still managed to hold the door open for others. After lunch one day, Mary and Gabby cleaned all the desks with Lysol wipes. One morning, Mark set up everyone's chairs. Zeth handed out the math textbooks when he returned from recess. When Mary knocked over her own geometry set and tissue box during a test, Gabby tidied it, having completed her test first. Terry did not directly prompt these actions, although she acknowledged and reinforced them with comments, such as: "That would be helpful, thank you"; "Thank you for thinking ahead"; "Isn't that so nice of you to hold the door open until everyone has passed through"; and "Very helpful of you Gabby. Mary, did you thank Gabby?" Public acknowledgment of desirable and exemplary behaviours is a means of moral instruction, identified by *The Manner in Teaching Project* (Fenstermacher, 2001; Richardson & Fenstermacher, 2000, 2001) as *showcasing*.

There were, nonetheless, moments of disappointment when Terry's expectations were not met and when some students were generally less helpful than others. On one occasion, Terry requested that Alexander and Zeth accompany Noah to the office five minutes before recess was to end. When the time came, the two boys ran away instead. While on crutches, Heather asked Kay to bring her a laptop from down the hall. Kay refused. Heather asked a second time. Kay refused a second time but got one for herself. With a shake of her head, Terry told me, "Kay tends to do only what Kay wants. Sometimes Kay is very considerate, particularly related to issues of social justice, which she enjoys bringing to the class' attention. Not always on a personal level, though". Terry reflected, more philosophically:

> They are generally ready to crawl outside of themselves and their own, personal needs. And this is a good time to push it. But they do it at different times, are ready for it at different times. Hopefully by June they're well on their way to thinking about other people and then they continue that journey at their own pace.

FOSTERING RELATIONSHIPS WITH STUDENTS

Noddings (2002, 2008, 2010) advocates for teachers and students to engage in reciprocally caring relationships. This enables teachers to fulfill three morally desirable goals: address students' individual needs; facilitate students' communication of these needs; and provide a context in which a range of moral values are expressed by teachers and students, including, among others, care, compassion, kindness, respect, empathy, trust, thoughtfulness, acceptance, and sensitivity. Residual relationships between Terry and former students were visible within the first weeks of school. On recess duty, Terry was often surrounded by a group of older girls with whom she easily chatted and laughed. Several regularly visited the classroom to say hello. Terry explained how deeply connected she felt to them, forecasting her intentions with the present class:

> They're former students. That's why, when they graduate in grade-eight, I get all teary eyed, too, because look at where they're at, and look at how far they've gone from when I had them in grade-four, and look at what they're doing. I get really connected to them. Sometimes they come back. So, for example, the grade-fives have a connection with me and Wendy if they're still doing peacemakers. In grade-six, they may still have a connection to me if they play field hockey. And then they just keep coming up and

talking to me. But by grade-seven or -eight, they may not even say hi any-more. At graduation, when I look at them sitting facing the audience, and we stand up and applaud them, sometimes they look at me. Last year, I remember there was a girl in the first row. She barely said hi to me any-more by the time she was in grade-seven or -eight, but at graduation I can see her looking at me. And I'm giving her the thumbs-up. I started crying because she's come a long way. I had her in grade-three, and I had her in grade-four. Another girl has graduated, and she's actually come back once already this year to say hi. So that's nice. She visited me.

It appeared to be girls, not boys, who continued to connect with Terry beyond their year together. This is likely a result of girls being more inclined toward relationships than boys (Gilligan, 1982), rather than a bias on Terry's part. The discussion above, regarding Terry's desire to give equal attention to all students, supports this supposition. The following reflections indicate that Terry does make a personal connection with boys:

I feel good when somebody does something, and I can say, "I had that person in grade-four". I feel good about that, and I say, "Look how far they've gotten, and wow, look at what they're into now". So I'm proud that way, just seeing how far they've come. I think if Zeth continues to make progress socially and emotionally, the way that he is making prog-ress, I think I'll feel most proud about him and where he's come from. I'm really tough on him, so honestly, I'll be most proud of him.

Observing Terry with former students alerted me to her early efforts for fos-tering relationships with the present class. Two related strategies emerged—gathering information on each student and informally interacting with each student. Terry began to gather information prior to the first day of school. She read comprehensive reports prepared by former homeroom teachers and con-tacted the teachers directly to clarify points of confusion or gain additional insight. Of particular interest were students' academic profiles, including best and worst subjects, and their personal and social circumstances, including home life, friendships, and social difficulties. Some of the more fun facts were shared with the class on the first day of school to help foster relationships among them, as noted above. Information gathering continued on the sec-ond day of school, with an activity called Tell Me All about Yourself. Stu-dents completed a questionnaire identifying their preferences on a variety of topics, including books, music, places visited, hobbies, school subjects, and aspirations and reflected on what happiness, friendliness, and caring mean to them. Although most relevant to her vision, Terry believed the reflections were too sophisticated for the start of grade-four and intended to return to them later in the school year. She never did, however. This symbolizes Terry's general lack of enthusiasm for prepackaged activities relating to nonacademic curricula. While this activity was intended to help Terry know her students, something she prioritizes, its applicability was limited. The information Terry gained from interacting with the students soon made any potential information from a worksheet superfluous.

Informal interactions were embedded in the everyday proceedings of classroom life. For example, Terry welcomed the students at the door or in the hallway most mornings. This allowed for small exchanges, updates, comments, anecdotes, and hugs. She ate lunch with the class three of the five school days per week. Although Terry often sat at her own desk, she chatted and joked with the students. Indoor recess, when weather was inclement, was similarly spent with students, rather than engaging in professional work. During special activity days, Terry wandered with groups of students through the displays and events. Finally, while on recess duty, Terry watched the students play, rather than chat with other teachers. These behaviours were intentional: "I want the students to know that I'm there for them. I hope it comes across to them that I'm not, in any way, so above them that I'm not approachable". Watson (2003) confirms that eating lunch with students affords teachers the opportunity to build caring relationships with them. It is reasonable to assume that this opportunity is also embedded in the other daily interactions between Terry and the students. Terry's comments, in regard to helping students with problems, support this assumption:

> I might say, "It's about you and me, and it's just about that question" or "It's just about this problem you're having, and it's nobody else's business". I would hope that it makes for a closer connection just with me, knowing that I'm paying attention to just them on this issue, that I'm giving my time to help them sort things out, and it's one-on-one time, rather than in front of everybody else.

Gathering information and informally interacting with students are not exclusive strategies in fostering caring relationships. Terry explains, "The more that you allow [students] to talk about things and the more that you listen to where they're coming from, the more you understand about who they are". Interactions provided Terry with information, and information enabled Terry to initiate interactions. This reciprocity is illustrated by the following examples of conversation openers, all of which led to a personal exchange between Terry and the students: "How was your sister's piano recital last night, Heather?"; "Mary, did you like how the book started off?"; "How was skiing, Alexander?"; "What did you think of the hockey game last night, Connor? I know you watched"; and "I took your grandma's advice, Zeth. I'm eating more quinoa. I bought the bread she recommended, too. It's yummy". In typical fashion, Terry worked to engage all of her students:

> There are some who don't speak a lot, who I don't get to know right away. For example, Connor and Zeth, they're so out there with their personalities it's really easy to read them. But the quieter ones like Gabby and

Alexander, and sometimes even Mary, those are the ones I have to make more effort to get to know because they're just in the back. They do their thing. Those are the ones that I remind myself to go and check in with.

Such comments acknowledge the manifestation of archetypal personalities. With years of experience teaching children this age, they serve as guidelines in the process of knowing students, interacting with them, and ultimately building relationships, but with a caveat:

It is about learning from the kids you've had in the past and from knowing them. I learned so much from handling different types of emotions, the anxiety piece. Zeth is a highly anxious kid. Had I not had a really anxious kid in the past I would not have been as calm or confident with Zeth. It takes a lot of time to learn about the children. I have also learned how to observe the common elements, so it's not as time consuming for me as it was maybe 10 years ago because of all the different kids that I've had. Although, sometimes I can be typecasting too quickly. But you learn both ways.

Facilitating the reciprocal quality of teacher-student relationships, Terry enabled the students to also know her and to also initiate interactions. She signalled the former on the first day of school by asking the class what they had heard about her and continued throughout the year to be open and honest with them. To encourage students to initiate interactions, Terry remained physically, intellectually, and psychologically accessible. Examples of this have previously been provided. Evidence that the students had, in fact, engaged in relationship with Terry and were reciprocating her caring feelings, overtly emerged toward the end of the first term. Drawings, letters, cards, and poems created by the class began to accumulate on the bulletin board by her desk. One particularly illuminating card, which all of the students signed, read as follows (corrected for spelling and grammar):

To Ms. Kennedy,
    You've been here for this year and hopefully the next, too. You stood here with us all year long. Together we are strong. And when you smile, the sunshine falls upon your face and brightens up the room. You've understood our problems and helped us work them out. For us, you are our world, our sunshine and moon. Just knowing that you will be there waiting for us when we get to school gets us up in the morning. And all we are trying to say is, although we have not been here long, you've taught us all we know. Together we are strong.
    Merry Christmas.

These sentiments express the students' love for Terry and the personal bond they feel with her, as well as hint at happiness and a sense of community.

The themes are not mutually exclusive—more likely inextricably and inevitably intertwined when teaching morally.

Teacher-Student Relationships as Moral Instruction

Teacher-student relationships, grounded in pro-social and moral values, enhance the moral messages teachers convey through other means, such as modelling and dialogue, and provide a context in which students might practice expressing a range of values (Noddings, 2008, 2010). These morally educative benefits are relevant to Terry's practice, as illustrated in examples provided throughout this chapter and the next. Additionally, such relationships provide a standard for the relationships Terry hopes to foster among her students in an effort to nurture a class community. She characterizes the relationships as caring, such that students are expected to care for each other as Terry cares for them. This is articulated, as follows:

> If someone is unhappy about being in school or someone is unhappy when they leave for the day, I need you to really step up. I need you to help. Even if you are scared that you might get in trouble, you still need to step up, whether taking responsibility or helping the person cheer up. It's not all about you. You need to take care of each other, to help me with that.

Terry's interpretation of "take care of each other" coincides with care ethics. Accordingly, Terry demonstrated the role of *carer* in meeting the needs of the *cared-for* (Noddings, 2008, 2010).

Terry provided opportunities for students to practice what she modelled by situationally upholding expectations for them to care for each other and, more formally, by placing them in the role of carer. The former is illustrated in an example of "showcasing" (Fenstermacher, 2001; Richardson & Fenstermacher, 2000, 2001). Terry announced:

> Noah did three things that were really kind. He asked Bonnie if he could use her desk before he just sat down at it. He got her another chair when she couldn't find hers. And he saw that Alexander was upset, and he moved over to him to help him out.

Kindness, respect, helpfulness, and compassion were directly referenced or implied as inherent in caring for others. The following anecdote recounts how similar expectations were reinforced with the girls:

> The weather held the promise of an early spring. Mr. Smithson took the girls outside for physical education class, to enjoy the sunshine and play a game of capture the flag, usually a favourite. When they returned to the class, Bonnie, Kathy, and Sammie were crying, and the rest of the

girls seemed to be in various stages of shock. Mr. Smithson followed on their heels and beckoned Terry from the hallway. When Terry returned a few moments later, it was lunchtime. Several students anticipated this, already unpacking their bags. Terry asked the boys to take their lunches next door to Wendy's room and asked the girls to remain with her. She unpacked her own lunch and sat at her desk in quiet thoughtfulness. Then she began talking.

TERRY: Everybody, heads up. Before we talk about what happened, a quick note. I don't want you to leave for March break feeling like this. It's not that bad; so let's just calm down. Who can explain? Someone who is neutral, not directly involved in this.

HEATHER: We were playing capture the flag and we quit. Then we saw Pia was crying.

FRANCES: Everyone was yelling and saying, "You're not my friend anymore". People weren't following the rules.

TERRY: What is at the root of all this? What is the main part of the problem?

HEATHER: We didn't want to play because they said we were cheating. They started screaming at us.

Several girls began to call out at once. Terry settled them down. Bonnie, Kathy, and Sammie continued to cry into their desks, leaving their lunches untouched.

TERRY: Why are these three girls so upset then?

No response.

TERRY: Is there one humungous issue? What could that be?

FRANCES: It started because Pia was crying.

HEATHER: I don't think it was about Pia crossing the line. I think it was about Pia being sad. It was about both teams yelling at each other.

TERRY: Do you really feel that you no longer like each other? Don't you still like each other?

FRANCES: Yes, but we always have trouble with capture the flag.

TERRY: What is it about that game that messes you up so much?

Several suggestions were offered, including, "The rules keep changing"; "People get involved when they don't have to"; "Some people are very competitive"; and "Some people take over". Terry quickly summarized:

This game makes you care more about winning than you care about each other. Can we agree then that only those involved in a situation will deal with that situation and that you won't be so competitive, that you'll care more about each other? It's a game that always brings out the worst part of your competitiveness. Let's cool off, and we can talk a bit more later.

While Terry did note particular negative behaviours, the central concept was caring for each other. Once the students assumed that responsibility, she hoped the antisocial behaviours would dissipate.

Terry did not forget the incident during the two-week break that followed. While dismissing the girls for their first physical education class upon returning to school, she reminded them of the problems last class and asked them to put each other first and to take care of each other. "No game is worth the kind of upset you experienced last class. Think of and try out some solutions for this. Let me know later how it all went and what worked". The girls returned from the class with smiles, proud to report that the game went well, and there was no fighting.

At times, the role of carer was formalized. For example, the learning buddies program paired Terry's students with Carol Lindsay's grade-two class. They met every week in the library for a craft, reading, or writing activity. Terry's students provided learning support for Carol's students during these activities. Terry instructed her students to be role models and mentors for their buddies and to support them more generally throughout the school year. This expectation was rigorously upheld when Noah kicked his learning buddy. Although such behaviour is unacceptable anytime and under any circumstances, Terry was particularly distressed because Noah had breached his position of carer in the relationship. "I am so disappointed because this is not the impression I had of you. I thought you were a great partner and that you are always so helpful to Irene", she said with tears. In a final example, Terry created partnerships for a science project, with consideration for the following needs: "Zeth needs someone calm"; "Heather needs someone with good ideas because she has trouble coming up with them herself"; "Can't let Alexander get distracted. Who can keep him on track?"; and "Frances is a very hard worker, but she works too fast and she is messy. Kathy will help keep Frances's work neater. Can Kathy put up with Frances's quirks, though?" A *carer–cared-for* relationship is implied, one in which some students will meet the needs of others with kindness, care, respect, and sensitivity, as Terry consistently models for them.

CONCLUSION

In attending to her own conduct, behaviours, and practices; a class community; and relationships with students, Terry focuses on moral values that underlie professional practice, classroom life, and the structures of schooling. These include, among others, fairness, equity, equality, honesty, respect, and trust as informing pedagogy, procedures, rituals, and routines; cooperation, helpfulness, kindness, and responsibility as informing chores, duties, and tasks; and empathy, sensitivity, care, and compassion as informing power, authority, and relationships. Applying this range of moral values embodies the notion of teaching morally. It ensures that Terry treats students and the academic work they produce as well as their time together at school with the upmost regard; that she and the students attend to each other's dignity, welfare, and well-being, and ability to flourish; and that the learning conditions are morally justifiable. These are worthy and desirable ends.

These ends may also serve other ends. By rendering expressions of moral values visible through modelling; by obligating students to be helpful, to take responsibility for themselves and others, and to care for each other; by demonstrating how to be in caring relationships; and by providing a variety of opportunities for students to practice expressing moral values, teaching morally assumes an educative quality in Terry's practice. Further, class community and caring relationships provide fertile macro and micro contexts, respectively, for the more direct methods of conveying morality. Thus, while the next chapter is focused on four such direct methods, Terry's practices for teaching morally, as discussed in this chapter, remain relevant.

REFERENCES

Campbell, E. (2003). *The ethical teacher*. Maidenhead: Open University Press McGraw-Hill.

Campbell, E. (2004). Ethical bases of moral agency in teaching. *Teachers and Teaching: Theory and Practice, 10*(4), 409–428.

Campbell, E., & Thiessen, D. (2000, April). *Moral and ethical exchanges in classrooms: Preliminary findings*. Paper presented at the Annual Meeting of the American Educational Research Association, New Orleans, LA.

Campbell, E., & Thiessen, D. (2001, April). *Perspectives on the ethical bases of moral agency in teaching*. Paper presented at the Annual Meeting of the American Educational Research Association, Seattle, WA.

Colnerud, G. (1997). Ethical conflicts in teaching. *Teaching and Teacher Education, 15*(6), 627–635.

Fallona, C. (2000). Manner in teaching: A study in observing and interpreting teachers' moral virtues. *Teaching and Teacher Education, 16*, 681–695.

Fenstermacher, G. D. (2001). On the concept of manner and its visibility in teaching practice. *Journal of curriculum studies, 33*(6), 639–653. doi: 10.1080/00220 270110049886.

Fenstermacher, G. D., Osguthorpe, R. D., & Sanger, M. N. (2009, Summer). Teaching morally and teaching morality. *Teacher Education Quarterly, 36*(3), 7–19. Retrieved February 25, 2010 from http://vnweb.hwwilsonweb.com.

Furman, G. C. (2004). The ethic of community. *Journal of Educational Administration, 42*(2), 215–235.

Gilligan, C. (1982). *In a different voice: Psychological theory and women's development*. Cambridge, MA: Harvard University Press.

Hansen, D. T. (1993). From role to person: The moral layeredness of classroom teaching. *American Educational Research Journal, 30*(4), 651–674. doi: 10.31 02/00028312030004651.

Jackson, P., Boostrom, R., & Hansen D. (1993). *The moral life of schools.* San Francisco: Jossey-Bass.

Johnston, D. K. (2006). *Education for a caring society: Classroom relationships and moral action.* New York: Teachers College Press.

Lickona, T. (1991). *Education for character: How our schools can teach respect and responsibility.* New York: Bantam Books.

Lickona, T. (2004). *Character matters: How to help our children develop good judgement, integrity, and other essential virtues.* New York: Touchstone.

Lickona, T., & Davidson, M. (2005). *Smart & good high schools: Integrating excellence and ethics for success in school, work, and beyond.* Cortland, NY, Washington, DC: Center for the 4th and 5th Rs/Character Education Partnership. Retrieved January 3, 2009, from http://www.cortland.edu/character/.

Noddings, N. (2002). *Educating moral people: A caring alternative to character education.* New York: Teachers College Press.

Noddings, N. (2008). Caring and moral education. In L. P. Nucci & D. Narváez (Eds.), *Handbook of moral and character education* (pp. 161–174). New York: Routledge.

Noddings, N. (2010). Moral education and caring. *Theory and Research in Education, 8*(2), 145–151. doi: 10.1177/1477878510368617.

Nucci, L. P. (2009). *Nice is not enough: Facilitating moral development.* Upper Saddle River, NJ: Pearson Education.

Oser, F. (1994). Moral perspectives on teaching. In L. Darling-Hammond (Ed.), *Review of research in education* (pp. 57–127). Washington, DC: American Educational Research Association. Retrieved April 16, 2008, from http://www.jstor.org/stable/1167382.

Oser, F. K., Althof, W., & Higgins-D'Alessandro, A. (2008). The Just Community approach to moral education: system change or individual change? *Journal of Moral Education, 37*(3), 395–415. doi: 10.1080/03057240802227551.

Osguthorpe, R. D. (2008). On the reasons we want teachers of good disposition and moral character. *Journal of Teacher Education, 59*(4), 288–299. doi: 10.1177/0022487108321377.

Richardson, V., & Fenstermacher, G. D. (2000). *The Manner in Teaching Project.* Retrieved July 29, 2008, from www.personal.umich.edu/~gfenster.

Richardson, V., & Fenstermacher, G. D. (2001). Manner in teaching: The study in four parts. *The Journal of Curriculum Studies, 33*(6), 631–637.

Ryan, K., & Bohlin, K. E. (1999). *Building character in schools: Practical ways to bring moral instruction to life.* San Francisco: Jossey-Bass.

Solomon, D., Watson, M., Battistich, V., Schaps, E., & Delucchi, K. (1996, December). Creating classrooms that students experience as communities. *American Journal of Community Psychology, 24*(6), 719–748. Retrieved November 38, 2009, from http://www.metapress.com.

Watson, M. (2003). *Learning to trust: Transforming difficult elementary classrooms through developmental discipline.* San Francisco: Jossey-Bass.

Wynne, E. A., & Ryan, K. (1997). *Reclaiming our schools: Teaching character, academics, and discipline.* Upper Saddle River, NJ: Prentice-Hall.

5   Teaching Morality
Practices That Are Overtly Instructive

Fenstermacher, Osguthorpe, and Sanger (2009) define teaching morality as conveying to another that which is good and right with the intention of affecting his or her character: "The teacher is providing to another person the means for becoming a good or righteous person" (p. 8). In accordance, the present chapter is organized around direct practices that Terry employed to advance the moral growth and development of her students. These include the remaining four of seven practices identified in Chapter Three—virtues instruction; morally laden discussions; disciplining with a goal of self-regulation; and service activities. Each is fully explored in a designated section of this chapter. A fifth section portrays how Terry assessed the students' moral learning in the absence of a formalized accountability process at the school.

This chapter is intentionally situated after having established that Terry teaches morally. Her approach to teaching morality is embedded in a context of teaching morally, and her identity as a moral educator is embedded in her sense of self as a moral person. The latter point is assumed in Campbell's (2003) two-pronged conception of moral agency, from which I draw my own understanding. Fenstermacher, Osguthorpe, and Sanger's (2009) definition of teaching morality similarly assumes that by providing the means for students to be good and righteous, teachers know what is good and right and are themselves good and righteous. Thus, discussions in the previous chapter on expressing moral values in conduct, behaviours, and practices; nurturing a class community; and fostering caring relationships with students underpin and are echoed in the discussions here.

VIRTUES INSTRUCTION

Virtues instruction refers to formal didactic lessons on particular virtues or moral values. Dating back to the inception of schooling, in early 17th-century Colonial America, this represents the first method of school-based moral education. At that time, it was situated in the religious scriptures of New England Calvinist Puritans and served to advance the Bible's moral code

(Balch, Saller, & Szolomicki, 1993; McClellan, 1999). The contemporary incarnation is secular, focusing on moral values that are objective and universally espoused. As such, *The Manner in Teaching Project* (Fenstermacher, 2001; Richardson & Fenstermacher, 2000, 2001) and The Moral Life of Schools Project (Jackson, Boostrom, & Hansen, 1993) empirically identified applications of this method in regard to both independent and academic curricula. Further, virtues instruction is the primary method associated with character education.

Terry's research and participation in creating Middlevale's Character Development Program deeply immersed her in character education literature, particularly the work of Thomas Lickona, and influenced her moral education practices to include virtues instruction. Accordingly, she delivered formal, preplanned lessons and imparted informal, spontaneously interjected messages, the latter of which I also include as virtues instruction, when focused on a particular moral value. Virtues lessons were easily observable and mostly derived from the school's character program, as well as the grade-four language arts and health curricula. The health curriculum, in particular, contained a unit called Character Education, which was organized around five themes: respect, responsibility, growing and changing, *what would you do?*, and peace seeking. Terry and Wendy did not follow this curriculum as laid out in the guide but used their discretion to create lessons from among the different topics and activities. Virtues messages, on the other hand, were less visible, occurring in the moment and often passing quickly. They usually did not reference a program or piece of curriculum but advanced an aspect of Terry's vision for classroom life or the students' growth, development, and well-being. These lessons and messages are presented anecdotally, arranged according to the core moral values being communicated—compassion, courage, cooperation, respect, responsibility, and inclusiveness, respectively. Compassion, courage, respect, and responsibility are school values. Respect and responsibility are also part of the health curriculum. Cooperation and inclusiveness, although informally reinforced throughout the school, more particularly underpin Terry's vision for a class community and enable her collaborative pedagogical approach to teaching and learning.

Compassion

> It was only the second day of school. The students sat at their desks quietly watching as Terry wrote five letters on the whiteboard: R, R, I, C, C. Tapping the board, she said, "Somewhere around this room are five very important words that start with this". Several students pointed to the poster behind Terry's chair and in unison recited: "respect, responsibility, integrity, compassion, courage". Terry joined in. Continuing, she said, "Tell me what you know about any of these words". A few ideas were offered. Terry listened and nodded at each but refrained from commenting

herself. With the exception of Zeth, who was new to the school, the students were quite familiar with the school's values and the rhetoric around them. When the class quieted, Terry mused aloud, "They're all really closely related . . . respect, courage, responsibility". She trailed off, hoping the students would have more to say. But no one did.

After a brief pause, Terry affixed a large pad of chart paper to the whiteboard and focused in. "Let's talk about what this word means, what it looks like. Give me your ideas". She had prewritten *compassion* at the top of the first page, intentionally prioritizing this moral value to jumpstart the formation of a class community.

Kathy raised a hand and replied, "You show it by helping people if they're hurt".

Bonnie added, "Being really nice".

Terry nodded in agreement with both and said, "Another word for compassion is being kind, or nice, as you say Bonnie". She looked around the room expectantly and then asked, "If something happens at recess, how can you show compassion?"

"You could let everyone who wants to play with you, play", Frances called out.

Terry took a moment to record their comments on the chart paper. Then she prompted the students further with an upcoming scenario. "Some of you are going to be trying out for sports teams. Some will make it on the teams, some won't. What could be compassionate in this situation?"

"You could say, 'You did a good job. Sorry you didn't make it'", Paige suggested.

Terry persisted, "And what if *you* got onto the team? How would your reaction help other people? When you get in, you feel happy and proud. But how could your reaction help your friend who didn't get in?" The students were silent and a little confused. "You see? There are two sides to this. Your reaction will either make them feel worse or better. So feel proud and happy, yes. Just keep others in mind. Be gracious about it if they didn't get in. Don't go whooping all over the place", Terry explained. She noted this on the page as good sportsmanship. Terry scanned the class one more time, waiting for further comments. But on this second day of school, none were forthcoming. It wouldn't be long, though, before the opinions of these students would be difficult to cap, happily so, in Terry's mind.

Figure 5.1 replicates the chart paper, as it looked at the end of this lesson. It remained posted on the whiteboard at the front of the class until the next virtues lesson, although Terry did not reference it further.

This lesson on compassion was delivered within the context of the school's character program. At the time, Terry did not make connections to classroom life, her vision for class community and student happiness, or

COMPASSION

– Helping people

– Being kind to others

– Including others

– Showing good sportsmanship

Figure 5.1 Compassion

her efforts to foster caring relationships with and among the students. These connections were made, however, through many informal and spontaneous messages on compassion and related moral values of empathy, care, sensitivity, and kindness. Several examples are noted throughout the book and are not repeated here.

Courage

Courage is a school value. Yet, the formal lesson on courage was not delivered within this context but, rather, as part of a language arts Risks and Consequences Unit:

> Students took turns reading aloud the story of a mouse, a crow, and a cat. The mother mouse, Mrs. Frisbee, was desperate to return home as quickly as possible with medicine for her sick son. She decided to take a shortcut, knowing that she might encounter the dreaded cat. Hurrying along, Mrs. Frisbee came across a crow, named Jeremy, who was caught in some wiring. He called out, "Can you help me?" Every second counted for her son's well-being and her own safety, but Mrs. Frisbee still stopped to untangle the crow. As feared, the cat appeared just as she was finishing and threatened them both. Grateful for Mrs. Frisbee's help, Jeremy said, "Hop on my back". She did, and he flew them both to safety.
>
> "So what do we learn from this story?" Terry asked as she closed her book. The students replied with various comments related to reciprocated helpfulness. Terry prompted them further, "But helping another when you risk yourself takes courage, don't you think?" Flipping over the compassion page of the chart paper, Terry revealed a fresh page with the word *courage* written at the top. "Can you give me some examples of courage from the story?" she continued. Terry recorded what the students called out, without comment or judgment. She then asked the

students to think about their own lives: "Besides the story now, what can we say about courage?"

With only a moment left of class time, Kathy offered, "To make friends with people from different countries or backgrounds". Terry recorded this final comment and dismissed the class.

Figure 5.2 illustrates how the chart paper appeared at the end of this lesson.

The lesson continued the following week with two guiding questions: When is a risk worth taking? How can concern for others help us to overcome our own fears? The students were asked to reflect on both questions in the context of the story and in the context of their own lives. Over the next few days, they also wrote a paragraph describing a risk they had personally taken and the consequences of taking that risk. Their paragraphs recounted positive outcomes, such as successfully trying scuba diving for the first time and taking home a particularly feisty dog from the kennel, which became a beloved family pet. Each was illustrated and posted above the students' lockers in the hallway as a badge of honour for their courage.

Terry also imparted informal and spontaneous courage messages. When individual students needed encouragement and support for presentations and performances, for example, Terry might say versions of "I know this

---

COURAGE

- Helping someone else when it puts you in danger

- Jeremy tried not to panic as the cat came closer

- Jeremy had the courage to put his trust in Mrs. Frisbee

- Mrs. Frisbee took the route with the cat

- Mrs. Frisbee went to get medicine and came home to her son

- Asking people for help

- To make friends with people from different countries or backgrounds

Figure 5.2   Courage

will take some courage". When reviewing vocabulary from a language arts story on runaway slaves, also part of the Risks and Consequences Unit, Terry explained the word *asserted* as follows: "You are assertive when you stick up for a friend and say, 'Stop bothering my friend. She is not happy about it'". Although Terry did not articulate the word courage, she illustrated courageous behaviour. In neither of these situations, however, did Terry reference the poster on her wall that read *You'll always miss 100% of the shots you don't take* or the formal lesson she had previously delivered.

Courage messages were more ardently communicated when one of the boys in Wendy's class called Paige "fat". Terry and Wendy facilitated the following discussion with the girls from both classes. The boys were sent to Jonathan Blakely, Assistant Head of School, who is in charge of campus discipline and the character program, and Tom Sinclair, the school psychologist, for a separate discussion on respect.

WENDY'S STUDENT:	At my old school, my friends made up this thing where you just say "thank you" if someone insults you. I like that.
WENDY:	Is that what your teachers taught you at your old school?
WENDY'S STUDENT:	Yes, and not to let it bother you.
TERRY:	We aren't saying that it won't bother you or not to let it bother you. You can't let people think that it didn't hurt. You need to tell them that it isn't okay. Say, "We were having fun, but that wasn't cool. That wasn't fun". You need to have courage to do this. They need to know that it hurt your feelings or made you feel uncomfortable. You need to tell them.

Several girls raised their hands and shared additional stories related to similar incidents with boys. Terry reiterated her message.

TERRY:	You need to stand up to them and tell them. Boys and girls are different, and that's okay. But you have to be clear if you want them to understand you. Boys need to do that, too. Give them your message very clearly, even though that might be hard. If you need help with that then come to us, and we will help you.
WENDY:	We try to help each of you all the time, but we can't be there for everything. We try to teach you skills so you can deal with most things on your own as you get older. I hope with you being strong and maybe getting help from adults then it won't continue to happen.

Although the incident involving Paige that initiated the discussion was not directly referenced, Terry hoped that she would internalize the message, nonetheless:

> It is courage. I want Paige to stand up for herself, to be less agreeable. When we had that talk about the recess incident, name-calling, calling her "fat", I talked to her. I said, "It's not okay that he said that". And then we had the class conversation with Wendy about clearly communicating how you feel. Hopefully she's starting to want to develop courage to do that. I'm keeping an eye on it. So when we were doing seating, she said to Noah, "I'm always in the back". She didn't say it out loud to me, but I caught it. And I said, "Paige, say something if you don't want to be in the back. It's still your choice". But she just said, "I'm good". So I'm working on that to develop.

Cooperation

Cooperation and helpfulness are related in Terry's practice but not synonymously so. Terry distinguishes the two values by implying a reciprocal quality to cooperation that she does not necessarily attribute to helpfulness. In showcasing (Fenstermacher, 2001; Richardson & Fenstermacher, 2000, 2001) her own behaviour, for example, Terry tells the class, "See how Ms. Bell and I are working together? I'm reading the instructions while Ms. Bell is doing the experiment. That's how you work in groups". This is the core message of the following anecdote:

> "You will be working with other adults in whatever you choose. The more people you work with, the trickier it is. This is a way for me to see how we are working together", Terry explained just prior to the first group activity. The students organized themselves into their four house groups. This meant the groups would be uneven in number, with two, three, four, and five students. Each group was provided with a piece of Bristol board, precut into a circle. "Divide your circle into wedges, according to how many you are in the group. Each of you is responsible for creating a drawing for one wedge. But here's the thing, your drawing has to connect with the drawings of the neighbouring wedges. And the overall theme for your circle must relate to your particular house and the values that your house promotes. That means you will all reflect the value of respect, plus either compassion, courage, integrity, or responsibility". Terry paused before continuing. "Here's the most important part. You need to work together and plan this out. Discuss how you will do this. You have to talk it out. Talking is important. You have to cooperate with your peers and give them a voice, too". With these instructions, Terry layered her expectations for moral learning. The content of the activity required the students to reflect on the school's values. The process required them to practice cooperative behaviours and expressing related pro-social and moral values.

One of the groups set themselves up in the hallway; another went to the library; two remained in the classroom. Before long, bodies and supplies were strewn across the floor, circling Bristol boards. Terry floated among the groups, monitoring their progress and reinforcing her expectations with the following values-laden comments: "Don't shut down the ideas of others in your group"; "Offer up different ideas"; "How can you say those are hideous? That's not cooperation or compassion"; "Why can't he offer up his opinion?"; and "I like the way this group is cooperating and helping each other".

The largest group of five students, however, was not working well together. In agitation, Sammie ran over to get Terry. "Zeth is drawing our house as a burning building! None of us want him to do it, but he won't stop. He won't listen to us". Terry walked over to where the group was stationed.

"This is a group project. You all have an obligation to tell Zeth your objections clearly, and Zeth, you have to talk about all the ideas and listen. Talk it out before you start". Knowing this was not the first time the students were asked to do group work, or the first time they had heard about cooperation, Terry removed herself to let them try again. But Zeth continued to colour the house red and draw flames on its roof. The rest of the group sat by and watched, looking completely helpless and more than a little stunned. Terry returned and reiterated the same values, but this time more directly and forcibly. "What I see here is not a sharing of ideas. Everyone in the group needs to be happy. Everybody needs to be able to live with the decisions. I can see this is not the case. So how can you talk with each other and come to an answer? What questions might you ask when you have a problem with ideas? What can we do when some like the idea and others don't like it?" Terry paused for a moment so the students might have time to reflect. No one answered. She continued, "Here are some questions to ask: Why are you doing it? Why might some like this but others not like it? This project is about learning how to get along, and you have to be respectful of each other to do that. You also have to be responsible for what you're doing and saying. These circles will be hanging up and people will see them. Your parents will see them. Other members of our school community will see them. You all need to feel proud of what you create".

Zeth was the only one in the group who did not make eye contact with Terry, but he did stop colouring. This small sign of progress pleased Terry. Although her comments were addressed to everyone, she was aware that the problem stemmed from Zeth's behaviour. She did not want to target him directly, however, so early in the school year. He was having several difficulties integrating into the new school environment. "All of them are against him. They get it. But he needs to also get the point", she later reflected privately with me. Over the course of a week, and to varying degrees of craftsmanship, the four groups finished their circles and hung them from the clothesline for parents to see on curriculum night.

Terry determined that she needed to more formally reinforce cooperative behaviours among the students. The lesson described below was delivered to the grade-four girls during a health class while the boys were in physical education. The boys had a similar lesson later in the week:

"What is cooperation?" Terry asked.

The students variously called out, "Work together"; "Get along with other people"; and "You're not alone in this classroom".

"What does cooperation look like? Give some examples or a situation", Terry continued.

The students responded enthusiastically, "sharing stuff"; "playing with people"; "a game without fighting".

Terry flipped over the chart paper with the lesson on courage and began to record on a fresh page. "Good start. Working in pairs is easy. What about larger groups? Why is that harder?"

Kathy offered, "There's more people to share with".

"Yes", Terry agreed. "What about coming up with ideas? You have to talk and listen. There are more opinions. You have to be respectful. You have to compromise. Sometimes you can't have it your way. You have to give and take in a group situation. That's part of cooperation, too". The girls called out several additional comments, which Terry added to the list.

Figure 5.3 reproduces what was written on the chart paper at the end of the discussion.

COOPERATION

- Compromise
- Listen
- Be fair
- Sharing stuff
- Playing with people
- A game without fighting
- Being respectful
- Compassionate to others
- Having family meetings to dissuss
- Doing chores

Figure 5.3   Cooperation

This page remained posted in order to guide the group activity that immediately followed:

> "I would like to see cooperation", Terry announced. She randomly assigned the girls to five groups of three, and one group of four. Their task was to create a cooperation cube, such that each of the six sides reflected this value. The content and the process of the activity, once again, coincided. The girls drew elaborate images with the following themes: playing, singing, reading, and doing chores with others; and sharing, holding hands, taking turns, listening, and making new friends. Terry occasionally interjected messages regarding what their conduct should be. She asked, "Are we cooperating?" and "No one's bossing anyone around, right?" The cubes were hung on the clothesline for a week. When they were taken down, Terry asked the girls if she might keep them as a reminder of how nicely they had been prepared, inferring process as well as product.

Although focused on cooperation, Terry is aware that these lessons and messages also impart other values, including general values of negotiation, communication, and listening, as well as moral values of helpfulness, respect, and responsibility. She explains:

> It's a negotiation, rather than controlled by just an individual. And then, if there are difficulties in the partnership, we would hope that they would be able to learn to negotiate, with maybe some help from us. There was one particular group of three who had a lot of trouble working together, and there were a lot of arguments. And that also is valuable because from that you take away if things aren't working well; how are you supposed to work through conflicts? So putting them in groups helps to develop better communication and negotiation skills—learning to listen to the other person's ideas and knowing that the other person has as much value as you. Therefore, you have to say, "Okay, what's your idea?" And then working through and making sure that your idea and their idea are in the project, not just your idea. So that's basically what we're trying to work through. Respect, remaining respectful even though you recognize there are differences, and then making sure that you put in your share of the responsibilities.

Respect

Terry delivered two formal lessons on respect and informally imparted several related messages. This first lesson derived from the health curriculum. It was delivered to the boys:

TERRY:	In what ways can you see professional sports players showing respect?
WENDY'S STUDENT:	Shaking hands at the end of a game.
WENDY'S STUDENT:	If a player falls, sometimes players help them up.

TERRY:	So, it doesn't have to be someone you are friends with?
CONNOR:	If someone gets hurt at recess, you help them.
TERRY:	How about at art camp? Is it the same?
NOAH:	If someone makes something, say, "Nice sculpture".
WENDY'S STUDENT:	If someone says, "I'm really bad at this", say, "No you aren't"—because everyone makes a really big effort, and that makes them more confident.
TERRY:	Does it mean that if you're good at soccer you'll be good at everything else? Or if you are good at art, will you be good at everything else?
	Several boys responded in unison, "No".
TERRY:	So, if you are a really good musician, people may tell you, "You are really good". But if it is writing, you may have trouble with it, and someone can say they will help you. That is showing respect?

Terry paused for comments. None were forthcoming, so she continued.

TERRY:	Here's another situation. If you are at the theatre and watching a movie, or watching a play, how are you going to show respect?
CONNOR:	Don't tell what happens if you have already seen it.
WENDY'S STUDENT:	Don't throw popcorn at their face if you don't like the character. I saw that once.
NOAH:	You still need to clap.
CONNOR:	You don't make noises if you don't like it. Keep your negative comments to yourself.

Terry validated these answers with a nod, and probed further.

TERRY:	When the play has started, what do they ask of the audience?
WENDY'S STUDENT:	Turn off cell phones, and no talking.
TERRY:	Sometimes the ushers don't let you back to your seat right away when you go to the washroom. Why is that?
WENDY'S STUDENT:	It's annoying.
TERRY:	Sometimes it is also distracting to the actors.
WENDY'S STUDENT:	If you see how in a magic show the magician appears, like you see the door, don't go up and say, "Hey look. Here's how it happened".

The boys then fell silent, having exhausted their thoughts for the moment. Terry rose from the edge of her desk, where she had been perched, and handed around a two-page worksheet entitled How Do You Show Respect? (Schwartz, 1997). She began to read the first page with the class: "You are showing respect when you take care of your clothes, books, games, and other possessions; treat everyone fairly and kindly; care for your environment by recycling; listen when others are talking, and do not interrupt".

Two of the boys, however, had been doing just that for the past several minutes, calling out comments and generally being disruptive. Terry ignored them for a bit, hoping their behaviour would settle. When it didn't, she responded, "I'm sorry; I was talking right now. You are not being respectful of me or of each other". How ironic, I thought, that Terry should have to deliver this message in the mists of a lesson on respect. And yet, how helpful to have an authentic example. The rest of the session proceeded uneventfully. The boys worked independently on the worksheets, filling in answers to "Describe what the word respect means" and "Describe a time when you were especially respectful to someone or when someone was respectful to you".

Terry did not associate this lesson with any of the incidents that prompted its earlier-than-scheduled delivery, and its generality raises doubts that the boys made the connection themselves, particularly when two of them did not make a connection to their immediate behaviour. Further, this was the only virtue lesson for which Terry used prepared materials. As was the case with the Tell Me All about Yourself activity, summarized in Chapter Four, she admitted to not reviewing the worksheets once they had been submitted. This substantiates the earlier claim that Terry does not find prepackaged materials to be useful tools for moral education.

The second lesson on respect took place one week later. It is described anecdotally as follows:

"Frances has something very important and interesting to share with us", Terry announced as the students settled into their seats. Frances moved to the front of the class and began to read a letter from her grandmother that outlined the difficulties women of the Suffrage Movement endured, such as police arrest and physical abuse, while protesting for their rights. The class applauded politely when she finished. Frances returned to her seat, folded the letter, and put it away in her cubby.

Terry rose from her chair and moved to the front of the class, saying, "Last week there was something I was concerned about. It was comments made by the boys to the girls, and by the girls to the boys—bad comments about the opposite gender. I heard someone say, 'Wouldn't it be better if we had an all boys' school or an all boys' class?' And 'All boys are smarter than the girls'. That just isn't true. You have to be very careful of what you say. I don't want you to think that it is okay to say those things, because it isn't. It's not respectful to talk like this". The class remained quiet. After a brief pause, Terry referenced Frances's letter: "When you make those general comments about girls, it is similar to the prejudice that women faced. It is a form of prejudice, where you are judging the person simply because they are female. It is also prejudiced to say general things about all boys". Terry let a few moments pass in silence before dismissing the class for lunch.

The message had not been internalized, however, and the unacceptable gender-biased comments continued. A week later and without preface, Terry once again addressed the class on this issue:

"What was I saying about respect the other day? I had very specific things I was trying to get across". Terry was angry, and the class knew it. "Here are some ideas. Don't say bad things about other people. Don't insult anyone, especially the opposite gender. Hands are for greeting and not for hurting. You don't have to be best friends with everyone, but you do have to be polite. Remember?" No one dared to speak. Terry continued a little more gently, knowing that she had their attention: "I know it's really easy when you are a girl to be annoyed by boys and if you are a boy to be annoyed by girls. But what I'm trying to tell you is you have to try really hard not to fall into the same habits that other people may be into. If other people are saying bad things about girls, then you shouldn't join in. And if you are saying those things, then you shouldn't be. Today I heard that someone in this class said 'boys suck', and no one stopped them. You need to stay away from that language. If you say that all girls or all boys are bad, you know it's not true. When you talk to each other, individually, you know it isn't all bad. Sometimes you are helping each other. That's not bad. I know it's really hard, but it won't always be hard. We do need to be kinder to each other".

Terry had been standing at the front of the class. She now moved toward her desk and perched on its edge more causally. "So, let's deal with one thing at a time. When we do things to annoy other people, it may be amusing. We all do it maybe to siblings or best friends. We do it because it can be fun. You just have to know where to stop. You girls have to gauge the reaction of the boys. Look at their facial reactions. If they are getting more and more frustrated, then you know you've gone over the line, and you have to stop. You really need to watch their faces. And you also have to know, if you are on the other end, that you have to say 'stop'. And that is hard because you don't want to look like a party pooper or that you're not cool or that you're not fitting in. You have to be strong and courageous enough to say 'that isn't fun anymore'. That's the first part. Let's work on that".

Although these messages were contextualized as respect, they also include courage, helpfulness, and kindness, as they echo previous discussions separately conducted with the boys and girls. In addition, Terry noted vices of prejudice and discrimination as being inappropriate and intolerable.

Responsibility

Responsibility for others was discussed in Chapter Four as a learning outcome of nurturing class community. In the lesson and messages reported here, Terry focused on self-responsibility, primarily for one's behaviours and conduct. This became a priority in the third term of school, although related messages were imparted earlier:

> Terry began, "Over the past two weeks we have had issues. We've had issues at recess. And then last week there were issues with learning buddies. The buddies weren't happy with us. They were kind of disappointed. And then today at assembly Mr. Patrick read a poem. All of these things can fit under responsibility. How can these fit under responsibility?"
>
> "Saying sorry?" Heather suggested. Terry asked her to clarify in a sentence. "Say sorry when you hurt somebody", she replied. Terry recorded this on a fresh piece of chart paper.
>
> "This can also be a physical hurt", Terry reminded them.
>
> Connor called out, "When you do something wrong, don't run away from it".
>
> "Yes", replied Terry. "Lying just makes things worse. It makes it harder when you have to admit to it later. Nor do we like to hear at the grade-four level, 'I don't remember what happened'".
>
> "Sometimes we have that problem at peacemakers. When you have problems with the little kids and they say, 'I didn't do that'. We say, 'Tell the truth or you will just get in more trouble'", Kathy added.
>
> The class remained silent while Terry caught up with her recording. Then she asked, "What else?" No one replied. "Okay, now we have to look at the big picture", she continued. "Why do you think I leave responsibility for the end of the year? At the beginning of the year, we talked about compassion and courage. I have a specific reason why I chose this as the last one we do".
>
> "Because you wanted us to know that we have to commit to what we are doing", Frances suggested.
>
> Terry pressed, "That's an all year thing. What is it about the third term?"
>
> Kathy tried, "The talent show comes around and really bad things can happen".
>
> "Two people have mentioned this now", Terry said. "Same with track and field, right?"

Sammie added, "There's also responsibility for your speech".

"Okay. All good answers. But the end of the year means what?" Terry scanned her students with a smile.

"Grade-five", Connor yelled.

"So what is it about the end of the year and going on to grade-five that has to do with responsibility?" she asked Connor.

"You have to have responsibility", he replied.

Terry added, "Also, at the end of the year you can say to me better 'here are my responsibilities' because you've had the whole year to understand what that means. At the end of the year, I'm expecting you to handle your work a lot more independently. You're not going to need as much guidance as you did in September. I'm putting more of the responsibility on you to figure out how to progress. I know that in grade-five the teachers are going to expect that you are able to do certain things a lot more on your own than I expected of you in September. That's what I'm preparing you for".

Figure 5.4 depicts Terry's recording of this lesson.

RESPONSIBILITY

- Say sorry when you hurt somebody

- Admitting that you did something right away is better than avoiding the issue or lying (e.g., peacemakers with little kids)

- You have to stay committed to your clubs or sports

- Responsibility to do with speech work, academic work in general

Figure 5.4   Responsibility

Self-responsibility was also communicated by Terry informally and spontaneously, as follows:

> Two things happened this week that worry me. Two conflicts between students in this class. I said at the beginning of school that what you say and what you do will cause a reaction. Consequences will be good or bad. In grade-four we are asking you to be a lot more responsible for what you say and what you do.
>
> Alexander and Noah, I let you have those toys trusting that you would bring them out only at appropriate times. If they are a distraction you need to figure out where they should be. This is about you taking responsibility and me being able to trust you.

Lastly, in the following brief exchange, Terry had an opportunity to both showcase (Fenstermacher, 2001; Richardson & Fenstermacher, 2000, 2001) Noah's responsible behaviour and relay a corrective message to Connor:

> TERRY: Noah has chosen his spot in the line well. He has put himself at the end where he won't be tempted to talk.
> CONNOR: Good for him.
> TERRY: Yes. He is a role model for you, not an opportunity for you to be a smart aleck. Noah has chosen well. You need to decide, too, what is best for you and what is best for the class. If you are somewhere where you don't think you can control your talking, then you need to make a better choice. I want you to get to the point where you can do that.

These examples focused on the moral aspects of self-responsibility, as they further Terry's vision for the students' growth and development. There is, nonetheless, an implied connection with responsibility for others, in the context of a class community. This was made explicit when Terry announced, "I like how Alexander is just going about his business, along with our routines. That's very nice of you, Alexander. Makes life easier for the rest of us".

Inclusiveness

Inclusiveness is a defining moral value of Terry's vision for class community. Related messages and lessons were imparted in the positive, as inclusion, and in the negative, as exclusion. The lesson below occurred spontaneously:

> Terry allowed the students to pick their own partners for the reading session. "Remember that we are uneven in number today. Some group will need to have three people. Take in the extra person. Don't

exclude, because you all know how that feels to be excluded". The students seemed to have difficulty applying this concept. Terry noticed that Frances was reading on her own: "Frances, do you want to work by yourself?"

"No", Frances replied.

Terry asked, "Did you try to join a group?"

"Yes".

"What happened?" Terry pressed.

"They told me no", Frances said, choking back tears.

"That is disturbing. Can you please come with me and we can talk with them".

Kay, Pia, and Sammie had been reading aloud, as a group of three, but became quiet when they saw Terry and Frances approach. "Frances is feeling very unhappy. Can you explain to me why that might be?" Terry asked.

Understanding immediately what this was about, Pia replied, "We were already a group of three, and we didn't know that we were allowed to work as a group of four".

"But you didn't ask me. Frances was just excluded by you without any discussion. That much you do know about. What do you think we can do about this?"

Pia continued to speak for the others, "She can work with us".

"Can we also apologize to Frances? And next time, could you please ask me before excluding anyone?" The girls readily agreed, having meant no harm, and moved over to make room for Frances to sit. A similar situation had also occurred among the boys. It became apparent to Terry that the students understood they could be a group of two or three but not a group of four. The message of exclusion and inclusion seemed to be lost in this logistical detail.

At the end of the session, Terry called the class together for a chat. "That was an interesting test for me. Usually I make myself in charge of putting you in groups. How do you think it went today?"

Connor, whose group formed without incident, replied, "Fine".

Terry looked to the others, "In the beginning, was there some difficulty? I originally said pairs or groups of three. That may have been a problem. But some things were said to other students that weren't the best choice. What did we learn about resolving issues?"

Sammie answered, "Next time we'll ask if we are allowed to have four".

Terry continued, "Some people were unhappy. If you make someone unhappy, you have to resolve it so we are all happy. I saw some really touching experiences. People really stepped up to help and make them feel better. Cheer each other up so everyone goes home happy. Don't walk away. Please be kind to each other". Including herself, Terry noted, "We'll learn from our mistakes".

Terry was objecting to how students turned away their peers once the group quota had been met, without helping to ensure everyone had a group to work with. Although this lacked sensitivity, care, compassion, helpfulness, empathy, and responsibility, the students did not have antisocial intentions. They were simply trying to observe the school convention of doing what one is told.

This second example extends Terry's expectation for inclusiveness beyond the class community:

> It was the beginning of the second week in December, the last week of school before the break. Most of the students had elaborate vacation plans. Alexander was going skiing somewhere in Europe. Sammie's and Pia's families were going to India, together. Mark was returning to Australia to visit family. Everyone was looking forward to it. Kay, Zeth, and Connor had already left. Paige and Noah were sick and stayed home to recover. The class seemed quiet and empty. Terry carried on, however. This was the day her students would be making holiday greeting cards for their learning buddies. Realizing that neither Kay's, Zeth's, Connor's, Paige's, nor Noah's buddy would get a card, Terry announced, "If some buddies don't get a card, they will feel left out. Their feelings will be hurt. Please make sure that we make cards for *all* the buddies, if someone is away today. If you know they celebrate other holidays besides Christmas, you can make their card for that holiday instead".

By identifying a harmful outcome should a student not receive a card, Terry designates this act as moral in nature (Nucci, 2009).

Final Remark

Formal, preplanned virtues lessons were observed to be Terry's most unnatural means of conveying morality. She explains why:

> I don't lecture them about this too too often because then it feels like "here we go again—here's the compassion; here's the courage; here's the respect". It isn't connected to anything this way.

This viewpoint coincides with Terry's lack of enthusiasm for prepackaged learning programs and materials and her reasons for rejecting a moral education curriculum for the school. Terry prefers, instead, to impart virtue messages, spontaneously and informally, as they become relevant to classroom life, individual student needs, and the academic program. Compassion, courage, cooperation, respect, responsibility, and inclusiveness are examples, therefore, and not an exhaustive list of virtues conveyed by this method.

DISCUSSIONS

Noddings (2008) notes that "talk of some kind" (p. 169) is recommended in all forms of moral education. In the context of care ethics, *talk* is referred to as *dialogue* and is considered to be "the most fundamental component of moral education from the care perspective" (Noddings, 2008, p. 169). Dialogue engages participants in the mutual search for understanding, such that outcomes are not predetermined but emerge from a reciprocal process of speaking, listening, and reflecting (Noddings, 2008). In the context of cognitive development theory, *talk* is referred to as *discussion*. Discussion is morally educative when it encourages new thinking on moral and social issues: "Students hear differing perspectives and points of view, and experience challenges to their own positions coming from peers as well as the teacher" (Nucci, 2009, p. 102). Despite different theoretical foundations, dialogue and discussion manifested similarly in Terry's practice, and both educative outcomes were expressed in her rationale for utilizing this method of moral education:

> When you talk things out loud, ideas become clearer and you get practice in expressing yourself very clearly, and people get a chance to listen to you and understand who you are a little bit better. There's such value in being able to say the ideas out loud and hearing other people say, "Well, no, you don't understand". No one is encountering your views if it all stays in your head. So when you're being challenged by other people, especially your peers at this age, there's some meaning. You are forced to actually reason out much better.

While the terms *dialogue* and *discussion* might be used interchangeably in regard to Terry's practice, I prefer the term *discussion*, as it includes talk among groups of students and the entire class, not only between two parties, which dialogue implies.

Terry intentionally seized and created opportunities for discussions in the context of school and classroom life, social dynamics among students, and the academic program:

> There are some things that we do that are more naturally geared to developing students' moral judgment, moral reasoning, ability to process moral matters, ability to identify moral issues. So whether it's a newspaper article that talks about something like that and we briefly discuss it, if they want to. Then sometimes it comes up in stories that we read or recess issues that happen—so day-to-day stuff as well as some curriculum materials lead to those discussions more naturally, I think.

Four such discussions are recounted below. The first occurred spontaneously, while the students were being dismissed for the day:

"Please put this letter into your agenda, and give it to your parents tonight. It's from Mr. Patrick", Terry announced as the class lined up.

"Ms. Kennedy, what's lice? It says here that someone has it", Connor asked.

Terry replied, "They are little bugs that like to live in hair. They don't cause harm, but they are a bother, and they need to be treated with a special shampoo that gets rid of them".

"Why do we need a letter about it?" Mary pressed.

"Because lice travel from one head to another easily, and we can all get it pretty quickly. This way everyone can be watching out, and we can hopefully prevent that from happening", Terry explained.

As Terry had hoped to avoid, several children began to ask at once, "Who has it? Who is it?"

Bonnie speculated, "I think it is Amy because she isn't here". Amy was one of Wendy's students.

Kathy concurred but wondered, "Why would she need to miss school?"

"Why do you think the letter doesn't say who it is?" Terry asked over their voices.

"Because when they come back to school people will treat them different and won't want to play with them", Frances suggested.

"Right", Terry said. "Does it matter who has it?"

"Not really", Alexander offered tentatively. He had seemed as curious as the others but was starting to sense that it wouldn't be right.

Terry nodded, "They have a right to their privacy, don't you think? How would you feel if it was you? Would you want others to know?" All eyes were on Terry, but no one answered, understanding this to be a rhetorical question. "Make sure you think about treating others like you would want to be treated if it were you. That is a great guideline for a lot of things. We talk about that a lot, don't we? All we need to know now is that we should check our own heads to make sure that if we have lice, too, we get the treatment so it doesn't spread. There is no need to know anything else about it. And, if one of you does happen to find out who it is, I hope you will keep it private and not spread that around. That would be gossip, and it could be hurtful to the person. Do you all understand that?" Everyone nodded. The issue was not brought up again, and I never found out who in the grade had lice.

This discussion was not open-ended. Terry used a series of leading questions to guide students toward the morally correct course of action. In doing so, she conveyed messages related to sensitivity and respect, and evoked empathy by referencing the Golden Rule.

Students often solicited Terry's help to solve problems of a personal or social nature. The second anecdote depicts one such occasion:

> The June talent show is a really big deal, for the girls in particular. They begin in February to organize themselves into groups and to plan their song and dance routines. It is a potential hotbed of conflict, for which Terry braces every year. Kathy and Frances pair off early and decide to do a musical number. They ask Bonnie to join them, making the group a threesome. A couple of weeks go by without incident. Then, suddenly, Kathy and Bonnie inform Frances that they have lost interest in what they were planning to do and do not want to do it anymore. "Besides", they tell her, "we are in other groups, as well, and busy with those performances".
>
> Frances, however, is not involved with any other group. In tears, she approaches Terry for help: "I have no one to be with".
>
> "Bonnie and Kathy, can you please come to my desk for a few minutes?" Terry requested. "I hear there is an issue about the talent show, and not everyone is feeling happy. I want to know what is going on".
>
> "Bonnie and I don't want to do what Frances wants to do", Kathy begins confidently.
>
> "So have you talked about it? Have you talked about options?" Terry asks.
>
> Hesitating slightly, Kathy rephrases, "Well, we don't want to do it. And she wants to do it".
>
> "You just quit me", Frances says bluntly.
>
> Bonnie responds, "But we are too busy now. I have another dance and so does Kathy".
>
> "So that is a different issue. Let's take turns and tell what happened", Terry directs them with a familiar script. Each girl recounts her version. Occasionally, Terry reminds them to not interrupt: "You will be next". Frances speaks last and tells Kathy and Bonnie that she now has nothing to do for the talent show, and no one will let her join a group because they are already too far along in planning their performances. Bonnie and Kathy, both sensitive girls, falter as they hear this.
>
> "Do you each understand what the others are saying now?" Terry asks. The three girls nod silently. Terry continues, "So tell me, Kathy and Bonnie, how does it make you feel to hear Frances's side of things?" Kathy and Bonnie had been struggling to hold back tears and now cry openly. Terry turns back to Frances, "Do you understand things differently now that you've heard Kathy and Bonnie?" Frances replies that she does. "So what I'm understanding is that this is a matter of exclusion, even though you might not have intended it to be that way. I know you two girls would never do that intentionally to Frances. But we do

have to fix this. You have left Frances with nothing and no one for the talent show". Terry lets Bonnie and Kathy settle a little before continuing. "It's okay. That's what happens when we talk openly about problems. It's a good thing. We can't solve this right now, though. Think about possible solutions, all three of you, and let's keep the conversation open. Maybe by tomorrow we can have something decided that will be good for everyone". With these final words, Terry sends the three girls into the hallway for a private chat. "Come back when you feel you are ready". A few moments later the three girls return. Their eyes are still red, but they are holding hands and smiling.

The next day, Kathy and Bonnie tell Frances that they are willing to do something with her. However, Frances has been able to join another group, after all, and is feeling much better. "So now we need to think of some strategies for the next time there is a conflict like this one, so no one's feelings are hurt. We have not always been carefully considering our decisions and choices. We need to do a better job of that", Terry counsels.

Kathy offers, "We could ask for a private conversation with each other".

Bonnie adds, "We could walk away to cool down". Terry smiles and nods at the girls and then sends them out to join their peers for recess.

With this discussion, Terry focused on process. A particular resolution was not predetermined but allowed to emerge: "It's not our business to put forth what our judgement is and what our solution is. The thing is to listen to them and to ask them questions, and see if they can come forward with the solutions", Terry notes.

This third example involved a former class and is recounted in Terry's voice:

So, two strongly opposing views is always interesting. Going to the zoo we had an issue whether or not the zoo is the right thing. Someone believed that you shouldn't keep animals in cages and within small areas. She didn't want to go to the zoo, didn't think it was right. And then others said, "Well some of those animals are endangered and having them breed is helping to keep the species going". It was really good to have a debate because that's where you get the meaty stuff. We try to debate and respect each other's opinions. It doesn't mean that you have to change the other person's opinion. But I like the students to see how you can talk about two opposing views and still relate to the other person well. It's not about changing their minds. It's just being able to speak their minds freely, still being able to listen to each other. To open their minds up to different things and to being more open about differences.

Terry noted the potential for two morally educative outcomes from this one discussion—encouraging moral behaviours of respect and tolerance, particularly as the core ideas were opposing, and stimulating moral reasoning skills with exposure to different arguments.

Lastly, the current events unit of study was a substantive source of class-wide discussions. Each student was responsible for reading and presenting four magazine or newspaper articles of their choosing. Terry instructed the students: "You can take anything from sports. You can take anything from *National Geographic for Kids* about animals, about endangered environment stuff. You can take whatever you're interested in". Presentations were followed by clarifying questions and a discussion to generate greater understanding. Often the issues were of a moral and ethical nature. One particular series of discussions is recounted in the service activities section below. Although not repeated here, those discussions are also relevant for conveying moral values of justice, empathy, courage, compassion, sensitivity, tolerance, responsibility, and helpfulness, as well as encouraging students to deeply consider the moral nature of issues related to human rights, social justice, and the environment.

Because Terry utilized discussions to impart moral messages and convey moral values, as well as to facilitate moral reasoning skills, she was not always neutral regarding her own opinions:

> Sometimes it isn't right for me to give my own opinions or my own judgement because it might just be my perception of things. However, having lived a longer life than [the students] and knowing more things about the world, I feel in most cases I need to give them that extra, a different point of view.

Her decision to participate was often made in the moment, according to the particular issues and views raised by the students and how the moral core was being addressed. In the zoo discussion, two opposing views on the ethics of keeping animals were presented. Terry did not express her own opinion or challenge the students because both views were interpretations of respecting animals and, thus, the moral core was not in dispute. At other times, as with the lice discussion, Terry felt compelled to direct the students toward the morally justifiable position. She defended this in relation to her education goals:

> When I'm not comfortable with them having a certain perception, then I want to change it. Usually I try to respect everybody's opinion. But when there's something that I don't agree with and I can't leave it, then I have to give them the right direction or a different way to think about it. Otherwise, I wouldn't be teaching them anything. They'd just stay with the same thinking, and that wouldn't be right either. If it's something that is so blatantly wrong, then I would definitely say, "I don't think that what you just said can be said like that".

Finally, Terry may not endeavour to change the students' opinions, only to present an alternative that had not been raised. In a discussion about Afghanistan, for example, Kay referenced the moral value of helpfulness. Terry presented, as an alternative, the value of autonomy:

> Kay couldn't understand why Afghani people didn't want Canadian and American soldiers there to help them. She couldn't understand why in Afghanistan people would want to hurt Canadian or American soldiers. And I said, "It's okay. You don't understand that. It's okay not to understand that and to have your own opinion. However, they want to be able to solve their own problems, and they don't want foreigners in their land solving them for them". She still didn't get it, but at least there's another way to look at it.

As with other aspects of Terry's practice, there is a limit to the discussions she encourages. Those that are political in nature, for example, are discontinued or repositioned as moral. Terry illustrates the latter using gay rights as an example:

> If the issue had been with whether or not to give the same rights to gay people, I know where I stand with that. But in that particular case, because it's political, I wouldn't tell them that this is the right way. However, what I would say is, "You have to keep an open mind. And you have to make sure that people have equal rights or human rights. And are you treating them like human beings who are equal to you if you are saying certain things about them?" Or, "You want them to have the same rights as you". So then it becomes, "You can have your own opinions about that kind of stuff. But just make sure you see it as a way of dealing with human beings equally".

In addition, Terry believes that too much talk time can, at times, be unconstructive. She offers two examples:

> We had a tough time last week with the girls. And we're still having a bit of an issue with the girls because they're starting to get on each other's nerves. And so then they would come up and talk to me, in a foursome, about other people and what other people are doing that they don't like. The ganging up feeling is so uncomfortable for me. They said to me that they needed to talk. But on the other hand, at these moments when I know that things are happening, it's almost a cycle where if I do give them a lot of time and attention to continue to talk about it and to focus on it, then they feed on that, as well.
>
> And Zeth right now is very anxious about the concert again because he hates too many people looking at him. But I know not to bother saying anything about it to him and trying not to focus on it with him.

When he said to me, "Now it's the concert coming up again", I just said, "Yup. But you did it last time, and you didn't die". He may say, "Yes I did". But I don't talk about it at length with him because I think he has a tendency to think and think and think about it, mull it over till it makes him crazy with anxiety. Better to just get on with other things.

The negative potential of discussions is an interesting insight that contributes to Oser's (1994) concern regarding the morally negative outcomes or side effects of teachers' otherwise well-meaning pedagogies.

DISCIPLINE

Terry understands discipline as an educative process that helps students learn to make good and right choices and decisions regarding their own behaviours and conduct. As such, discipline is related to self-control and self-regulation, and is properly termed self-discipline. Terry illustrates this in the following two passages:

> More and more, that's where I'm heading. Part of being independent and being responsible is knowing yourself and becoming more aware of what your needs are. So if you think, Okay, I'm really close to my friend and I'm chatting too much, I want them to get to the point where they can think it through and say, "I need to get work done. So maybe this is not the best choice. I can talk to my friend at recess. But this is not the right choice for now".
>
> A boy sees a very good friend doing a certain thing; it affects the boy and he wants to do the same thing, too. And as long as the role model is a good role model, then I say nothing and I keep going. If the person is not a good role model, then I would take the follower and talk to him or her. I might say, "I know you are very good friends, and I am not questioning the friendship. I think it's very good that you are good friends. I just want you to realize that sometimes your friend isn't going to make good decisions. And in that moment when your friend isn't making good decisions, I want you to know yourself, within yourself, whether or not it's the right decision for you, too. Do not do it just because your friend is doing it". I don't break them up or anything, but I am making the follower more aware of the times when he or she could break away from doing the same things—that it's okay to do that and still be friends with the person but not necessarily do everything the other person's doing.

As with other aspects of her practice, Terry does not approach discipline from a theoretical perspective. Nonetheless, the beliefs and assumptions embedded in her sentiments are consistent with developmental discipline theory (Watson, 2003, 2008), as previously discussed in relation to school level discipline.

Three types of discipline practices can be discerned, which roughly align with Watson's (2008) indirect, proactive, and desist control techniques. The first relates to preestablished and broadly applied guidelines for desirable behaviours and caveats for undesirable behaviours. This typically implies conventional rules. Although rules are ubiquitously endorsed in moral education literature (Hildebrandt & Zan, 2008; Jackson, Boostrom, & Hansen, 1993; Lickona, 2004; Richardson & Fenstermacher, 2000, 2001; Watson, 2008), Terry admitted to struggling with their utility:

> I don't know if there are any hard-set rules in this class, aside from the fact that if someone's talking you have to listen. I don't do rules. Teachers are always saying that it's really good. And there are a lot of books that say it's really good to have your class come up with things that they are agreeing to for the entire year, like a student pledge. And that the teacher should do a pledge, as well. But I never find that I refer to it beyond that. It just never worked for me.

Terry's beliefs regarding rules are similar to her beliefs regarding assigned duties. When prescribed, both risk obfuscating or undermining students' obligations to reflect on their actions and make morally good and right choices and decisions. Terry asserts that in the absence of rules the students learn to behave according to what they "should do, rather than rules that should be followed". Therefore, moral values of respect, helpfulness, kindness, and care serve in place of rules to guide student behaviours. Accordingly, the rule that Terry identified in the passage above—"if someone's talking you have to listen"—was actually articulated to the students in terms of respect and helpfulness, as follows: "When she's talking, you are remaining respectful and listening"; "Everyone who speaks should be heard. That's being respectful"; "I think you have forgotten that when someone is talking you need to listen. Even if you aren't interested, you must listen. That's respect"; and "It is not helpful for anyone if you talk while others are talking. So if it is not helpful behaviour, please don't do it. It is probably causing someone a problem".

The second type of discipline practice, also proactive, is situation specific. Terry anticipates the potential for misbehaviour or behavioural confusion prior to field trips, assemblies, guest visits, and special activity days and coaches the students: "If I give them a 'Here's what I'm looking for'; 'Here's what it's going to be like'; 'Here's what I'm expecting', they're okay. They get it. There are no surprises, and they know what they have to do", she explains. When an author visited the school, for example, Terry briefed her class with the following message:

> It is a privilege to have a real-life author visit us. Take paper for an autograph if you want, but we should first find out if it would be okay to ask for an autograph. Don't just shove the paper in the author's face.

Good Morning!

This is rehearsal day so please be

patient and expect changes in our

schedule.

SMILE!

Figure 5.5   Rehearsal message

Although Terry did not name particular moral values, she identified behaviour that was considerate, respectful, and sensitive. A second example is depicted in Figure 5.5. This message was written on the whiteboard the morning of a concert rehearsal. Terry named the moral value characterizing the behaviours she was expecting but did not identify particular behaviours. Students were obligated to interpret for themselves what was meant by patience and what patience might look like in that particular situation.

Lastly, reactive discipline practices were used during or following undesirable behaviour. Terry's immediate goal was to stop the behaviour. Yet, her reprimands were morally instructive, coinciding with Watson's (2008) guidelines—namely, (a) addressing the cause of the problem; (b) prioritizing solutions, not consequences or punishments; (c) highlighting harmful effects; (d) engaging students in problem-solving and making restitution; (e) maximizing student autonomy and minimizing adult power; (f) assuming the best student motives consistent with the facts; and (g) focusing on the behaviour, not the individual. This is illustrated with two anecdotes:

> The behaviour of the grade-four boys during music class had not been good for a couple of weeks. In consultation with Terry and Wendy, the music teacher asked Tom to sit in and provide her with support. After

the second session, Tom returned to Terry's classroom with the students and asked Terry if he might have a few words with them. Terry, of course, agreed and sat at her desk giving Tom the front of room. Leaning against the stool, he began, "I want Ms. Kennedy to hear this, too. What I just witnessed about your behaviour in music was not good. If I was someone who didn't know you, I would think that you were a group of rude children. Now I know that isn't true. I know you are not at all rude kids. You are sweet kids. And Ms. Jacobs knows that, too. That is why she is very patient with you. I wanted to tell you my thoughts on this. I think that I owe it to you to let you know". Terry thanked Tom for taking the time to be with them in music and for talking with them now. With a smile and a wave, he left the class.

Terry closed the door behind Tom and assumed his position by the stool. "That's the second time a teacher has told me about something that's happened in music. I didn't say anything the first time because I thought you might be able to fix it on your own. I wanted to give you that chance. If you are responsible for this behaviour, please ask yourself, 'Are you being influenced by somebody else?' If the answer is yes, you are still responsible for your behaviour. So you may need to move somewhere else. In that case, you need to tell Ms. Jacobs that you need to move. Make a decision about this before you get into trouble. You are old enough to decide this". Terry paused and looked at each face before continuing: "The next time you have music, not tomorrow but the day after, you owe Ms. Jacobs an apology. I will walk down with you, and we need to apologize for what happened. My hope is that you will not need another talking to by Mr. Sinclair. I know you. You are excellent with me. I understand that you have other kids in music, as well, and that means lots of distractions. Now you need to step up and do better. I don't want anyone else telling me that you are not behaving".

In this monologue, Terry focused on behaviour, not individual students. She addressed the cause of the problem by suggesting it might relate to the influence of students from the other class. Rather than punishment, Terry proposed a solution. By requiring students to initiate the solution, she maximized student autonomy and minimized adult power. Insisting on an apology, Terry engaged students in making restitution. Finally, Terry assumed the best motives consistent with the facts by indicating that the students were capable of meeting her expectations for good behaviour.

The second example involves a group of students misbehaving while working outside of the classroom:

TERRY: This group needs to come back into the classroom. There is too much fooling around, and you are too noisy.

CONNOR: It wasn't all of us Ms. Kennedy. It was Sammie. She was acting silly and not cooperating.

SAMMIE: No. It wasn't me.
CONNOR: Yes, it was Sammie! You know it was. We were trying to
          work. We told you to stop a lot of times.
TERRY: Sammie, we've been through this before. You need to tell
          me the truth. You need to tell me what your part in all
          of this was. You need to take responsibility if you were
          involved.
SAMMIE: I wasn't doing anything.
CONNOR: Yes, you were. And you know it. We all know it!
TERRY: Okay. Thank you, Connor. Please take the group and set
          up somewhere in here.

The group returned to the classroom and settled around an unoccu-
pied desk. Terry continued privately with Sammie. "Now Sammie, if
what Connor is saying is true, and I think it is because the rest of the
group is nodding in agreement with him, you are accountable to your
group for your behaviour. You are preventing them from getting the job
done. You are wasting their work time. You need to take responsibility
for this and make it right. I am expecting you to do this now". Sammie began to protest, but Terry simply said, "Make it right, Sammie".
Then she walked away to check on another group. Sammie rejoined her
group. She did not say anything further by way of apology but began to
work with them on the project.

Terry targeted Sammie, holding her personally responsible for her behaviour and accountable to her group. Yet, by focusing on behaviour and not
character and by assuming the best possible motives she preserved Sammie's dignity. Terry also noted harmful consequences related to hindering
the group's ability to accomplish the assigned task. Instead of punishing
Sammie, however, Terry encouraged her to find a solution and make
restitution.

As moral education, these three types of discipline practices signal Terry's
expectations for right and good conduct and behaviour, both generally and in
particular situations; hold students to a moral standard; encourage students
to assess choices and make decisions regarding behaviours and consequences;
and provide feedback and opportunities for students to practice and improve
when problems arise and expectations are not met. Accordingly, Terry does
not aim to manage or control student behaviours and classroom life but,
rather, to facilitate the students' social-moral growth and development so
they may manage and control themselves, both individually and collectively.

SERVICE ACTIVITIES

In justifying the service experiences she provides for her students, Terry
alluded to the two morally positive outcomes claimed by advocates of service

learning—furthering students' moral learning, growth, and development and meeting global or local community needs (Hart, Matsuba, & Atkins, 2008; Howard, 2005; Howard, Berkowitz, & Schaeffer, 2004; Lickona, 1991; Ryan & Bohlin, 1999; Schuitema, ten Dam, & Veugelers, 2008):

> I love taking myself and children's minds out beyond what they are comfortable with. And to be just really aware of what's happening in the world and relate it back to their lives. To be helpful in a global way . . . I want to contribute to making the world a better place by going through the younger generation, and I'm hoping that I'll be able to plant some seeds to help them create a better world at some point in the future.

Consequently, Terry's students were engaged in a variety of activities, emerging from the outreach program Terry coordinated, the academic program, a cocurricular program, and students' personal enterprises. These are itemized in Table 5.1 and described below as adult-initiated and student-initiated, and action-oriented and learning-oriented. The distinction between adult-initiated and student-initiated activities was determined by who assumed and maintained the leadership role, teachers or students, respectively. The distinction between action-oriented and learning-oriented activities was determined by who benefitted most, recipients of service or students, respectively. This latter distinction is not always clear. In acting for the benefit of others one does have an opportunity to learn, and in learning, one may be motivated to act. This categorization, however contrived, does enable me to identify key insights regarding Terry's use of service activities to convey morality.

Adult-Initiated Service

Adult-initiated service activities were generally school-wide, with limited opportunity for student input, choice, or leadership. There were seven such

Table 5.1   Service activities

	Action-oriented	Learning-oriented
Adult-initiated	Terry Fox run *loonie* days Goodwill drive holiday hampers Global Village craft kits	special activity days (Remembrance Day, cultural diversity, and environmental sustainability)
Student-initiated	Pia's Tim Horton's pitch Kathy's Girl Guide cookies	current events unit

activities during my year of fieldwork—special activity days (SADs), the Terry Fox run, *loonie* days, Goodwill drive, holiday hampers, Global Village, and an end-of-year grade-by-grade project. As introduced in Chapter Three, SADs centred on the themes of Remembrance Day, cultural diversity, and environmental sustainability. They were not linked to particular charities, and the students had no active role in serving others. Instead, they were learning-oriented, raising students' awareness of relevant social-moral issues and needs, locally and globally; exploring moral values of courage, empathy, compassion, justice, tolerance, and responsibility; and stimulating problem-solving, decision-making, and moral reasoning abilities. Although SADs were organized by Jonathan and set apart from classroom life, Terry seized two of these opportunities to support her moral education agenda. Remembrance Day began as follows:

November 11 did not begin like the other school days. The students arrived in top uniform, polished and pressed from head to toe. The teachers and staff were dressed in business attire, full suits for the men; suits, dresses, skirts with blouses or sweaters for the women, mostly in solid and dark colours. Terry wore a black sweater, black skirt, and black dress boots. A small, teal-coloured silk scarf around her neck, and a red poppy over her left breast provided the only colour. She typically did not wear makeup, and today was no exception.

The students were hushed in the hallway and hurriedly shuffled off to homeroom. The usual lingering, chattiness, and commotion were not tolerated today. In the classroom, the mood was also palpably different, sombre and serious as the students quietly read. At 8:45, Terry asked the students to put away their books. "In a few minutes we will be going to the assembly", she began. "How is today's assembly different from our other assemblies?"

Connor answered, "It's Remembrance Day".

"Yes. What does that mean?" Terry continued.

"It is about remembering the soldiers who fought in the war to keep Canada safe", replied Kay.

"So what does that mean about the mood of today's assembly? Do you think we will be cheering and laughing like we do at other assemblies?"

In unison, the students said, "No". For the next few minutes, Terry continued to proactively brief the students on more specific behavioural expectations: "You will need to sit still for a long time. Also, it's polite to look at the speakers while they are speaking, and listen carefully".

In the hallway, Terry's class merged with other classes heading to the assembly, each walking in single-file. Only the sound of echoed footsteps could be heard. The gymnasium was arranged differently, with rows of chairs lengthwise in front of a stage and lights dimmed. Terry led the class to a row near the middle and counted 16 chairs from the aisle, seating herself in the 16th chair, near centre stage. I sat on the aisle, and the students filed in between us. Settling quickly, they engaged in the hour-long program, which included a student's grandmother who discussed her wartime experiences as a young girl, and a slideshow of Canadian soldiers who had more recently lost their lives in Afghanistan.

After the assembly, the students returned to the classroom. Terry announced, "There is something that you can do, a contribution that you can make. It is important for us to learn, to remember, but also to contribute". The students passed around large sized postcards with the school's emblem. Terry instructed them to write a message and draw a picture for a war veteran, acknowledging his or her contribution to the country. This was eagerly undertaken, with some students producing more than one. Noah declared that he would write a card for his great uncle. Terry asked, "What can you tell us about his war experiences?" When Noah couldn't recall anything, she encouraged him to find out and report back to the class.

Just before lunch, Terry gathered the postcards and said, "The soldiers had the courage to help. They took responsibility to protect our rights and freedoms. This is about us having courage to help, too, not in quite the same way. It is about us taking responsibility. What can we do as a class to help others?" This was Terry's introduction to the end-of-year service project. Several girls offered suggestions for fundraising. "These are some very good ideas. It can be other things, too, not just raising money. Let's think about it a bit more and we'll talk again. We still have time".

Terry used this SAD to convey several moral messages, including respectful and considerate conduct and behaviour; sensitive and compassionate attitudes and ways of thinking; and thoughtful and responsible contributions and actions.

Terry's moral messages on cultural diversity day were derived from a book called *Selavi: A Haitian Story of Hope* (Landowne, 2004). This is the true story of a young boy named Selavi who lived on the streets of Port-au-Prince with other homeless children. Sharing food, resources, and companionship, and generally caring for each other, they overcome tremendous hardships to form Radyo Timoun, a radio station that continues to operate, giving voice to children. After reading the story aloud, Terry provided some context by briefly recounting Haiti's struggle for social justice and history of resisting injustices. Not wishing to dwell on political themes, however moral in nature, she refocused on the community that the Haitian children created, to relay moral messages of empathy, compassion, courage, helpfulness, and responsibility as they relate to the class community.

The remaining six adult-initiated activities were coordinated by Terry on behalf of the school as part of the Community Outreach Program. They obligated students and their families to take action in regard to particular charities or needs and, thus, are predominantly action-oriented. The first three—the Terry Fox run, *loonie* days, and the Goodwill drive—were announced to the students as being self-evident, without explanation or discussion. This is likely because they are annual occurrences. The Terry Fox National School Run Day (The Terry Fox Foundation, n.d.) took place in the fall, to support the Terry Fox Foundation for cancer research. Middlevale participated with schools across Canada. The students sought online sponsorship from friends and family and ran for an hour around the local track. *Loonie* days refers to the Canadian one dollar coin, which depicts a loon floating on the water. On 11 designated days in the school year, the students wore street clothes to school, rather than uniforms. They paid one loonie, each time, for this *privilege*. The school collected approximately 4,000 dollars, which they donated to three local charities. Finally, for two weeks in the winter, Goodwill bins sat on the porch of the school's administrative building. Students and their families donated clothes, toys, and household items. The bins were collected by Goodwill (n.d.), which sold the items to fund charitable projects related to work skills development for people with disabilities, youth at risk, and those with barriers to employment. There was no classroom component to these three activities beyond Terry's reminders regarding participation.

The holiday hamper drive was timed for Christmas, to support a local family in need through the holidays. Terry announced this project to her class at the beginning of December and posted a note on the whiteboard identifying the class's assigned family, a single mother with a teenage daughter and an 11-year-old son. The note also identified the family's item requests, which included particular nonperishable foods, clothes, toys, games, school supplies, and store gift cards. Terry asked the students to discuss possible contributions with their parents and with each other. Two weeks later, the fully stocked hamper was sealed and labelled. Terry arranged for hampers to be collected from classrooms around the school and delivered to the corresponding addresses. In January, she read aloud a thank you note from the family.

The Global Village was spontaneously organized to raise funds for Haiti in the aftermath of the devastating earthquake of 2010. Each class created and ran a booth to sell items that were made by them or purchased, or to host an event where admission was charged. The latter included, for example, a disco dance and story reading. Upon Terry and Wendy's suggestion, the two grade-four classes made and sold notecards. Supplies were purchased by Terry and Wendy with their art budget and donated by the students and their families. During the Global Village, students, teachers, and staff visited booths around the school, participating in activities and purchasing items. The grade-four notecards sold out quickly, for two dollars each. The entire event raised approximately 5,000 dollars, an achievement that was celebrated at a subsequent school-wide assembly. Terry reviewed with the class the particular charities that were collecting money for Haiti and noted that the school would decide which to support.

Lastly, the end-of-year project is a school-wide expectation to be implemented by each class or by grade. Terry intended to have the students determine their project. On Remembrance Day, they brainstormed several ideas. However, the logistics of these ideas were difficult to manage, as Terry explains:

> The ideal thing to do would be to talk about what kinds of initiatives they would like to do. But having said that, at this age it's really difficult when they're presented with something or when they're presented with the open "What do you want to do?" for them to come up with initiatives that are feasible. So usually what Wendy and I would do is either give them a choice or tell them what we're doing. And that's what we end up doing. But it's only a one-day thing. Whereas when they go to grade-five next year, they do their project for a longer period of time, collecting things for the homeless, that sort of thing.

Consequently, Wendy and Terry determined that their classes would make craft kits for children undergoing therapy for neurological injuries and disorders. Terry introduced the project and described paraplegia and quadriplegia, evoking empathetic sentiments. She asked the students to search their homes for unused art supplies or to purchase supplies for donation. A few weeks later, Kathy, Bonnie, Sammie, and Pia assembled kits with crayons, paper, scissors, stickers, and glue on behalf of the class. Terry delivered the kits from both classes to a local rehabilitation centre.

Student-Initiated Service

Three student-initiated service activities provided opportunities for student leadership and enabled individual students to choose whether they would participate. Two were action-oriented and linked to charities: Pia's Tim Horton's pitch and Kathy's Girl Guide cookies sale. With Terry's encouragement, Pia addressed the class regarding a Tim Horton's contest to raise money for cancer research. Family and friends had created a fund in memory of her uncle, who had passed away. She explained that pledges could be made on the Tim Horton's website, which Terry wrote on the whiteboard: www.timhortons.com. The students asked logistical questions, but Pia was unable answer. Although Terry tried to help, there remained a lack of clarity. It is uncertain if any of the students contributed to this fund. Shortly after, Kathy addressed the class to sell Girl Guide cookies. This time,

Terry briefed the students, as follows, regretting she had not done so for Pia's pitch:

> I just want to explain that whenever people at school talk to you about something that they would like you to support or donate to, I would like you to do all the research and ask lots of questions and find out about the charity so you can make a decision if this charity is one you would like to support. This isn't a competition with your friends. It's about what you think is worthwhile to support. It can be a very private thing. And it is something that you can decide. We won't be judging you.

Several questions followed relating to the Girl Guides organization. Most students bought a box of cookies, as did Terry. This may have been perceived, however, as a desirable purchase, rather than a worthwhile charitable contribution.

The third student-initiated activity was learning-oriented, involving discussions generated by the current events unit. At times, the students chose articles to present that enabled Terry to probe themes and values related to service. One day in February, four such articles were presented. Kathy's article explained the distribution of supply kits to those living on the streets. Terry asked, "Who is organizing this?" and "Is it just referring to our city?" This began a discussion on homelessness and whose responsibility it is to help those who are homeless. Pia's article profiled dogs that are trained to find buried survivors following a natural disaster. Terry asked, "What other situations might these dogs be helpful in?" This led to a general discussion about natural disasters, including the earthquake in Haiti. Bonnie's article described a garbage patch in the Pacific Ocean that was polluting the waters and endangering wildlife. Terry asked, "What is our obligation to our environment?" Various responses related to the theme of stewardship and cleanup. Finally, Kay's article advertised a new human rights museum. Terry asked Kay to explain human rights and then posed the following questions: "Are there other examples?"; "Do you think this happens in Canada?"; and "Where do we still have some problems?" In answer to the last question, and referencing her own article, Kathy suggested homelessness: "Once I saw a homeless person get kicked out of a grocery store because it said customers only, and he wasn't buying anything". Several students offered stories of their own encounters with the homeless. Terry allowed an unprecedented 20-minute discussion to ensue, which ended with the morally laden comment, "You never know about people's backgrounds. Try not to pass judgement on them".

In the midst of this discussion, Terry had an opportunity to spontaneously convey the moral value of respect. Kay kept referring to homeless people as *hobos*. Terry explained that this was a derogatory term and suggested that *homeless* was more respectful. But Kay continued to use the term *hobo*. Each time she did, Terry patiently reminded her. Eventually, the other students joined in, calling out less patiently, "Don't use that term!" In addition to respect, the broader discussion conveyed moral values of justice, empathy, courage, compassion, sensitivity, tolerance, responsibility, and helpfulness and encouraged students' to consider more deeply the moral underpinnings of issues related to human rights and the environment. Terry reflected on this session, with satisfaction:

> I was pleased because it was really nice to see those kinds of issues being brought up. That it's not just all about sports or buildings. It's nice to see those kinds of issues being talked about in the classroom and among children. I thought that was a really good day of really good articles that reflects who I am.

Service Activities as Moral Education

In identifying the service activities as adult-initiated, student-initiated, action-oriented, and learning-oriented, observations regarding the morally educative potentials of each can be made. Service learning theory claims that service activities are morally educative when they comprise two key qualities. The first relates to simultaneously meeting the needs of recipients and the greater community, as well as the growth and developmental needs of student servers (Billig, 2000, 2009; Furco & Root, 2010; Hinck & Brandell, 1999; Schuitema, ten Dam, & Veugelers, 2008). This is what Middlevale terms *social action* and *soulful action*, respectively. Second, service learning must entail high levels of student leadership (Billig, 2000, 2009; Lakin & Mahoney, 2006; Morgan & Streb, 2001). Morgan and Streb (2001) empirically demonstrated that "when students have real responsibilities, challenging tasks, helped to plan the project, and made important decisions, involvement in service learning projects had significant and substantive impacts on students' increases in self-concept, political engagement, and attitudes toward out-groups" (p. 166).

Yet, none of the ten activities in which Terry's class participated upheld both of these qualities. They focused on either social action or soulful action. Terry might have increased the learning opportunities of social action activities by encouraging students to reflect on how one determines who is in need of help; what one's obligation is to help; what forms help might assume; and how one makes decisions to participate. More specifically, the problems that each charitable contribution addresses may have been given wider context. For example, in relation to the Global Village, Terry might have explored how the living conditions of some people make them more vulnerable to natural disasters; in relation to the holiday hampers, how poverty undermines human dignity; and in relation to the craft kits, how children with disabilities create art. Some of the themes were addressed in current events discussions but without reference to these activities. Conversely, social action initiatives were indicated in the SADs and current events learning activities, particularly relating to social justice, the environment, poverty, and human rights. Any of these issues might have generated an end-of-year project for the class. None were pursued, however, despite Terry's claim that "the service part ties in nicely because if you're thinking about it then you should act upon it. You can't fully believe in something and not take the action part or not be involved somehow". Finally, student leadership was minimal, at best, in all service activities. Even those that were student-initiated failed to meet Morgan and Streb's (2001) criteria for student planning, decision-making, and implementation.

While the moral education outcomes of each initiative, on its own, might be questionable, taken together this is a comprehensive service program with a range of learning opportunities. Terry's students applied reasoning skills to explore moral and social issues involving the environment, poverty, disability, and human rights. They were exposed to different ways of expressing generosity, helpfulness, empathy, compassion, care, justice, responsibility, and courage, including giving time, items, and money and promoting or advocating for a particular cause. They participated in positive, motivating, and intrinsically rewarding experiences that served the needs of others. Further, coordination of school-wide and classroom-based activities delivers consistent values messages across the campus and enables scaffolded learning, as service opportunities and experiences build year-to-year. By grades seven and eight, students do assume more leadership responsibilities, and the service activities, themselves, more systematically integrate learning and action or soulful action and social action, respectively.

ASSESSING PROGRESS

Imbedded in these four direct practices for teaching morality is the notion of assessing students' moral learning, growth, and development. Since Hartshorne and May's (1930) Character Education Inquiry, many formal assessment instruments have been created. These are often quantitative in nature and aligned with either character education programs or cognitive development theory. Some are identified in Chapter Two. Terry is aware of these assessments from her research on moral education but generally rejects them as irrelevant to her practice:

> I saw different ways of doing this kind of assessment when I did the character education research, but sometimes it's too specific, and I can't see a teacher actually doing it or being able to work with it. . . . Quantitative really scares me because morality is so subjective, and it's hard for me to give a number or level to something that is not straightforward or objective.

As a result, and uncharacteristically, our conversations on assessment were more reflective than descriptive and more philosophical than practical, and entailed more questions than answers. This is best recounted in Terry's words:

> The assessment piece has always been the one that's missing. How do you do that properly? And I never had a clear idea anyway, even when I did the character education research. Before you can observe values with kids, you yourself have to have a good idea of what it's going to look like, what they would have to do to make you go "ah-ha". There has to be development. Until you know what that is, then it's hard to take notes because you don't know what you're looking for.
>
> It's almost like you have to do a checklist. Is this person becoming more helpful? But then it's so generalized. You can get so general about those words that they lose meaning. You'd have to be really anecdotal about things. Is this person showing more remorse? So, for example, when things are going badly and he did something wrong, is he going to lie about it, or is he going to admit it more quickly than in the past? It has to be based on what they do. But I don't know if I can make a checklist, a general checklist that will catch all of that. I think it would be helpful because then you'd know what to look for. But for every year you'd have to start with the baseline of behaviours. So-and-so does this kind of thing, so and so does this, this, this, and this, for each child. And then once a term or several times a term you go back to the base and say, what's he doing now?
>
> One year our report cards did include all of that. And we levelled them with "needs improvement", "satisfactory", or "very good". But then it gets tricky because all the times that you don't see something, how can you say that they don't have integrity, that you cannot see evidence of integrity, when there are things that you can't know about throughout the day or things at home that you don't know about? Values are something that people have from home, too. If you are talking about values and then trying to assess kids on that, then are you impinging on what should be home stuff? Then will parents really agree to how you're doing the assessment and what you're doing with the results?

Despite the practical and conceptual challenges Terry identified, she was, nonetheless, supportive of a process for assessing students' moral learning, growth, and development: "I think more than anything else it's really important to learn to see where they're at socially and emotionally and morally, rather than academically", she notes. Without the use of an evaluation instrument or formal assessment protocol, and in the absence of a recording process and accountability obligation, Terry used her discretion to monitor the students' moral progress. Relying primarily on instinct, gut feeling, personal moral wisdom, and teaching experience, she informally tracked three indicators—spontaneous expressions of morality, benchmarks and milestones, and response to tests and trials.

Spontaneous Expressions of Morality

Spontaneous expressions of morality are moral behaviours, moral reasoning, and moral sentiments that have not been directly motivated or encouraged by Terry. Terry's expectation that students will volunteer for chores and tasks enabled her to intentionally track behaviours of cooperation and helpfulness, as well as responsibility toward others:

> I don't keep track in any formal way whatsoever. But I do notice the ones who consistently, for example, volunteer to hand out stuff. There are regular volunteers. They like to do that. And there are some who don't. And once in a while I notice that so-and-so hasn't done anything in a while.

Terry's pedagogy afforded students a high degree of autonomy, necessitating self-regulation and behavioural choices related to respect, consideration, thoughtfulness, and honesty, among other moral values. In the following brief anecdote, Terry takes note of courage and integrity.

> Requiring a spool to build his project and knowing that none were left in the bin of supplies, Connor kept the spool he found on the floor. He hadn't realized that it had fallen from Paige's desk a few moments earlier, despite her quite agitated search for it. Watching this unfold from the back corner of the classroom, I debated whether to intervene. Quietly, I approached Connor and, in private, informed him that the spool had actually belonged to Paige, and she was looking for it. I did not tell Paige, however, that Connor had found her spool. Returning to my seat, I watched. A few moments later, Connor returned the spool to Paige, simply saying that he had found it, and unobtrusively redesigned his own structure without one. Later, I briefed Terry on what had taken place. "That's exactly what I'm hoping for. That they understand the right thing to do, and they have the courage to do it", she said.

Obligating students to attend to each other's happiness at school provided Terry with opportunities to track expressions of compassion, kindness, and care. When students made cards for one another—as Pia and Kathy did for Noah when his soccer ball ripped and the class did for Connor and Pia when they had been absent for several days—Terry commented, "They do show concern for each other. Perhaps more so than other classes".

These morally positive behaviours were weighed against morally nega-tive behaviours, such as complaining, fighting, and throwing tantrums. Terry made note of these, as well, for the lack of patience, kindness, thoughtfulness, and respect they demonstrated. Decreased incidents of negative behaviours, as indicated in the first passage below, and increased incidents of positive behav-iours, as in the second passage, signalled collective and/or individual progress:

> We had more trouble at the end of the term, last term, with problems at recess. They were really, really tired of each other. And a lot of tantrums happened. Now there seem to be fewer.
>
> In the boys' health class several years ago, we ended up talking about one particular boy who was making the others feel bullied and uncom-fortable. It didn't start out that way, and I didn't know at the beginning of the discussion that they were all talking about the same person. But they all knew who it was, and he knew that it was him, as well. He was there. But it was actually a very good discussion. They were not talking about it in any sort of venomous way or trying to gang up on him. And that's how I knew they were not just thinking about themselves any-more. They were saying things that were very reasonable and in a nice, matter-of-fact way, and they weren't mentioning names.

Terry was, nonetheless, hesitant to attribute such progress solely to the moral education she imparts in the classroom, appreciating the many moral influences in the lives of students.

Moral reasoning is associated with cognitive development theory and relates to how one understands, judges, and justifies issues of a moral nature (Howard, 2005). Terry recognized progress, for example, when stu-dents independently and successfully engaged in conflict resolution and peer mediation: "They needed me last term when they had an issue to maybe sit in once in a while on their discussions, and they're not asking me to sit in anymore. So that's progress", she explains. I observed this, as well. One day just prior to recess, Pia and Sammy approached me, independently, to complain about each other. Anticipating how Terry might handle this, I said to each, "You can either try and talk it out at recess or let recess be a cooldown time and come back afterward to talk it out. If those ideas don't work, then you should get Ms. Kennedy's help". The girls returned 20 minutes later, arm-in-arm and happy to report they had worked out their differences on their own. When I recounted this to Terry, she replied, "That's what we hope for". Progress is similarly recognized when students minimize the potential for conflict by anticipating the consequences of deci-sions and actions. Terry provided the following example:

> The end-of-year talent show is in June. The problems start with picking the people that you want to do your performance with. What happens

to the people who you don't choose? How are you going to avoid insulting them? Are you going to be in a fight? So Kay came up to me privately and she said, "If I wanted to just invite four girls, to ask them to perform something with me and to be part of the group, how do I go about that without causing bad feelings?" And I thought, That's really good because at least she's thought about it. She's thought about the consequences of asking in front of everybody. I like that because it tells me she understood the possibility of what could happen if she had just said it in the hallway, in the bathroom, when other people were around.

Lastly, spontaneously expressed moral sentiments also indicate progress. Terry's efforts toward moral education are particularly validated when the students echo her words, as she notes:

I would like to eventually hear my students say things that I have been saying to them for months. So if I hear it back from them, then I know that they've been listening. If I hear them say things to each other that come from me, with respect to how to be with each other and what they should do for each other, then that is progress.

Terry recalls, for example, Kathy admonishing her group for not assuming responsibility to discuss and share ideas and for not respecting her suggestions enough to comment on them: "I'm giving you all kinds of suggestions", she said, "and you aren't telling me anything". This echoes Terry's prior directives to "Offer up different ideas"; and "Talk about all the ideas and listen".

Benchmarks and Milestones

Terry's expectations regarding expressions of morality were judged against benchmarks and milestones. This involves appreciating the ebb and flow of the school year, the age-related developmental stage of the class, and the maturity levels of individual students. Regarding the ebb and flow of the school year, Terry learned when to expect social-moral progress or decline:

There's a huge difference socially and emotionally between where they left me in December before the break and when they came back. And it happens every year. The ability to handle more is there as soon as they come back from the break. . . . They always seem to grow up a little bit more in second term. It always starts out really well, and they seem able to do more difficult things in general—even the girls solving their own problems. So with the current events, I was able to go there and get into those types of issues. And then coming back from March break there's more ability again. By November we usually see a little bit of a dip in how they handle themselves. And then February usually is a big dip. They kind of relapse and have issues with each other.

Terry accepted typical periods of decline but attempted to preempt problems, nonetheless:

> They were having problems at recess at the very beginning of February. And I warned them. We talked about it. And I just said, "Listen, there are usually a significant amount of difficulties that arise in February". I told them that straight out. "Let's just catch it when it's early. February is a tough month, so beware". I tried to give them a heads up, although sometimes they just can't help it, developmentally.

The last comment references the social-moral developmental stage of the students, which Terry understands, as follows:

> At 9 years old, turning 10, they are less patient with kids who don't have the same interests. And then they need to learn that they have to still be polite and respectful about that. At 9 years old, they're also ready to really think about the world outside of themselves and extend their compassion that way. . . . At this age, 10, 11, 12, they're very much into justice and what is fair. And sometimes it's too black and white. So there are conversations that you can have with them, social and moral, that you couldn't before. It could be discussions that just pop up.

Teaching experience and research on moral education provided Terry with these broad insights by which she is able to manage expectations and make reasonable assessments of her students' moral progress. Still, Terry accepts and appreciates that each child matures at a different rate: "It may reach some individuals more than others, but that's okay if some kids get it more than others". Knowing the students well enabled Terry to sensitively customize her expectations, as she illustrates with Paige and Noah:

> Paige has always been thoughtful so there's not much change for her. Paige just has to stand up more for what she is really feeling and to be more open about what she really feels, instead of saying, "I'm okay and everything's good". Noah always seems to be just outside of the main circle of radar. And he still does silly things because he's a little bit less mature than the others.

Paige met Terry's expectation for thoughtfulness. Yet, Terry hoped to see progress regarding courage. Terry recognized Noah's social immaturity and, thus, was more patient with his behavioural antics. Desired moral progress, however, would involve increased self-regulation.

Tests and Trials

Periodically, Terry conducted spontaneous and informal tests and trials to assess the students' collective or individual moral progress. She expressed such intentions when introducing the first small group activity: "This is a way for me to see how we are working together". Reflecting afterward, she noted, "It was a good introduction for me to see how they are working together, to get a baseline that I can grow them from". For Terry, "working together" is moral in nature, necessitating expressions of cooperation, respect, consideration, helpfulness, and responsibility. Terry similarly utilized a change in the seating arrangement to assess the students' level of sensitivity and consideration:

> With Wednesday's decision to change desks, I wanted to have them decide. I just thought, "Let me just see if they can handle this". Just to see whether or not they were ready to consider the needs of others as well as themselves. If I set it up correctly, let's see if it happens. And I set it up, and I gave them the criteria, and they were able to do it.

Values of dependability, self-regulation, integrity, and responsibility were under observation when Terry allowed the class to walk unaccompanied through the courtyard to the French cabane: "I'd like to see you go to French without me today. I'll watch you from the window". Finally, Terry tested moral reasoning ability, as follows:

> Sometimes, when they ask me a question maybe about a problem they are having with another child, I put it back on them. I see what I get from them. And it may not take one answer. I may have to ask them again and again in different ways. And I just sort of see what answers I get. And if the thinking process is more logical, more mature, then it shows progress.

As with spontaneous expressions of morality, and benchmarks and milestones, notes were not taken, records were not kept, and progress reports were not issued.

Limitations

The rotary teaching system at Middlevale made it difficult for Terry to fully track her students' moral development. She expressed frustration with this:

> I try to talk with the rotary teachers regarding their comments or marks for the children on report cards. I'd like to know what kinds of things are happening in phys ed. And if I don't see it in the comments, it's really hard to catch up with the gym teacher on it. It's easier for the rotary teachers, when they see approximately 80 kids, to have similar comments for each. And so for me to get a sense of how each individual child is doing, it's really hard unless I have those conversations

beforehand. And that's frustrating because there's no central way to sort of catch up with them and say, "Okay, are they kicking the soccer ball properly? Are they cooperating when they play games?"

Terry attributed the problem to logistics. In the passage above, she references the number of students each rotary teacher assesses. In the passage below, she references the school's physical structure:

> I think it's hard for us to communicate with each other sometimes because we are in different buildings. We don't even see each other a lot. On this floor we see each other and become one community. Downstairs [is] another community. And then the other building is a community. That's what makes it hard to communicate with each other, to actually stay in touch.

A school-wide process for assessing student's social and moral development, linked to the Character Development Program and school values, might help to overcome these barriers and provide Terry with more consistent and insightful report card comments.

In addition, Terry resigned herself to the fact that the students' moral development was often not visible during their time with her. Consequently, she maintained a long-term perspective on assessment:

> If they had difficulties with maintaining healthy relationships with their peers in grade-four, and I cared enough to pay attention to them and give them strategies to keep working on it no matter how frustrating, and if that is still valuable two years later when the same issues come up again, and they can reflect back to what I've taught them or to some of the strategies that we've talked about, then that's good. . . . If they can remember that it was Ms. Kennedy's class where we did this and this and this in order to raise money for whatever, or we had a guest speaker who talked about other cultures and other people and how to think about other people, that is really rewarding for me.

Terry recalled with satisfaction the following specific example:

> Four kids took it upon themselves on the weekends to have a little lemonade stand. They went out in their community and actually handed out pamphlets and information. They went online and looked for the Free the Children website and copied samples of that and handed those out, too. So a talk that we would have in class affected them beyond class and actually continued to affect them in grade-five, where they wanted to have a cooking club and give the food that they made to a youth shelter.

Assessment of students' moral growth and development, therefore, was tempered by the humbling acknowledgement that Terry may only be planting seeds and may not witness how well these seeds grow and flourish.

The Results

Despite these limitations, assessment results were informative. When progress was noted, Terry credited the students, showcasing (Fenstermacher, 2001; Richardson & Fenstermacher, 2000, 2001), praising, and thanking them while still pressing forward toward a vision for their growth and development that remained just out of reach:

> The underlying thing is I always want to better the stage they are at. I want them to try and go a little bit beyond it each time. When I get a sense of their capabilities, I think let's try and up it a little bit.

Behaviours that were negative, reasoning that was morally questionable, benchmarks and milestones that were not met, and tests and trials that were unsuccessfully accomplished were opportunities for further teaching and learning. Several examples have already been recounted. Terry was cautious, however, to avoid creating a morally oppressive atmosphere, seeming dogmatic and absolute, oversaturating the classroom environment with morality, and generally desensitizing the students. She reflected on the latter:

> I think my feeling is, if I keep pointing out [morality] to them they'll just tune out when I do. It doesn't become meaningful anymore. I don't want to just throw it out uselessly, and then they just think that they're words and roll their eyes at me.

Accordingly, further teaching might not be explicitly moral. In the wake of bullying and exclusionary behaviours among the girls, for example, Terry delivered the following two lessons, one on feelings and emotions, and the other on social power:

> It had been a rough week, and I knew Terry was frustrated. "Why emotions?" I asked. "Why not a lesson on respect and care, for example, like Lickona would recommend?"
>
> Terry replied, "We're trying this for the first time". I watched with anticipation, as the boys went off to gym class, and Wendy's girls arrived for health. Without much of an introduction, Terry began a video about feelings and emotions. The girls watched quietly. When the video ended, Terry said, "Let's name some of the emotions from the video".
>
> The students called out: anger, sadness, happiness, anxious, depression, stress, guilt, and relief.
>
> Terry wrote each on the whiteboard.

TERRY: Any comments or thoughts? Can you connect any of these to your own experiences?

KATHY: Both of my dogs died.

TERRY: How did that make you feel?

KATHY: Really sad.

TERRY: Yes. I can imagine. What about anger? What has made you angry?

BONNIE: Sometimes I'm angry at a friend, and I want to say something mean. So I just walk away.

HEATHER: Sometimes my parents fight, and I have to leave the room. My sister yells at them to stop.

SAMMIE: I get stressed about math sometimes.

WENDY'S STUDENT: Ever since grade-one some people have been bullying me. I exploded. I got really angry and broke my door.

TERRY: The DVD showed some strategies for these things. It said that if you don't talk about it, it may explode on you, come out all at once. Does that scare you, the kinds of things that you can do or want to do when you explode with emotion?

No one replied.

TERRY: Okay, let's talk about some strategies to deal with emotions because you don't want to make a bad decision in how you act on your emotions.

FRANCES: When you are anxious, you can try to think of attitudes and actions that can help you.

TERRY: Can you give us an example?

FRANCES: When you want to jump into the deep end of the pool but are afraid, think of three bad things, like I'm going to drown, I'm going to bump on the bottom, and something gross is down there. Then you think of three good things.

TERRY: Let's write some strategies on the board. You are all peacemakers for the school so you can take ideas from there, too.

The girls called out several ideas, including "Talk and share your version of what happened"; "Write it down, journaling"; and "If talking doesn't work, walk away".

TERRY: We have to end here. We are never finished talking about this though. We're always talking about these things and trying to understand ourselves.

The essence of this lesson lies in the connection between emotions and morality. Yet, moral values of self-respect and respect for others, in relation to students' comments on anger, bullying, and yelling, remained intentionally implicit.

The following lesson on social power also occurred during a health session. The girls took turns reading aloud from a chapter called Bullies & Rule Setters in the book *A Smart Girl's Guide to Friendship Troubles: Dealing with Fights, Being Left Out & the Whole Popularity Thing* (American Girl Library Series, 2003). Terry instructed them to highlight particular passages as she commented on them and asked reflective questions. Below is a selection of these passages, interspersed with Terry's commentary:

> Your power lies in the part of you that's in charge—the strong, smart part of you. If someone is trying to take your power, you're being bullied. And if you're being bullied, you have to fight back. (p. 59)

> TERRY: This reminds you that you are strong enough, and you can take charge of the situation. Otherwise you leave the power with the other person, and you are not taking the power for yourself.

> One girl seems to have it all together. She knows how to make others feel good about themselves. (p. 60)

> TERRY: She sounds like a good role model, don't you think?

> The Rule Setter: She sets "the rules" and decides who's in and who's out of her group. (p. 61)

> TERRY: That sounds a bit like what is happening with you girls. Could the other girls do something when one of the girls sits with you, instead of just following the girl who gets up and leaves, for example? Who decides that this one girl should tell the others what to do, when to do it, and who to do it with?

In concluding, Terry did name some of the underlying moral values: "You need to trust your own strength, and you need to trust your friends and the friendships. Most of all, you need to have empathy and respect". She did not, however, link this discussion to the school's value of courage, which also applied.

Assessment results were not always obvious, as either positive or negative. For example, Terry originally assessed as negative, behaviours related to asking where particular students were when they were not in class. Upon further reflection, she decided the behaviours were well-meaning and, thus, essentially positive:

> I realized that they feel a little out of sorts when people are missing. I used to think it was because they just wanted to know. They were being busybodies. And I used to shut it down by saying, "It's none of your business" or, "Some things are private, and maybe you don't have to know where they are". But I think there's more to it than that. I think they just need to know where their people are. It's almost like an appendage that's missing. So now I just tell them ahead of time, before they ask me, "Where's Sammie?" [I say,] "Sammie is sick today"; "Kathy has gone on a dental appointment". Just so they know. I think that means they feel like a community. They're sort of feeling for each other. And they feel that something's not quite the same without everybody.

Rather than suppressing what she had believed were prying and meddling behaviours, Terry learned to reinforce the students' sensitivity and caring.

Further, not all negative assessments required a response. Sometimes Terry intentionally allowed the moment to pass. For example, when reading the holocaust novel *Number the Stars* (Lowry, 1989), the students did not recognize the moral significance of the story as previous classes had, despite the following prompts from Terry:

> CONNOR: I went to the museum.
> TERRY: Which museum was that, Connor?
> CONNOR: I think it was in Washington.
> TERRY: Oh, do you mean the holocaust museum?
> CONNOR: Yah, that's the one.
> TERRY: Tell us about it. What was it like?

Connor did not remember much, and Terry did not press him.

> KATHY: I know about Anne Frank.
> TERRY: Tell me about her.
> KATHY: She wrote a diary.
> TERRY: That's true. Have any of you written a diary?
> PIA: My sister does.
> TERRY: Is hers famous?
> PIA: No.
> TERRY: Kathy, can you tell the class why Anne Frank's diary is famous.
> KATHY: Because she died?

> TERRY: I think that has a lot to do with it, yes.
> CONNOR: And also she was hiding from the Nazis.
> TERRY: Right.
> KAY: Oh, yah. I know about her. She had to hide in the wall with other people, in a very small place.
> TERRY: Can you imagine what that would be like?

The class went silent. Terry waited, but no one had anything further to add. She continued reading from where they had left off, deciding not to raise the underlying moral issues. She explains:

> In some years, it feels like it's a class that you can really get into the moral discussions related to the holocaust. But other years, I feel that I have to wait until it comes from them before I go any further. If it's a matter of clarifying some information, like who were the resistance fighters, then I will do that because it's a fact. But I haven't felt like I wanted to push these guys. They just seem too young. They don't ask why; they don't ask how. They don't seem aware of the grey areas. I am through the bulk of the book, and I haven't had any of the discussions that I have had in previous years about who was responsible for this and why this all happened. But they still love the book.

Only rarely did Terry inadvertently miss an opportunity for teaching morality, replying to my queries with variations of "Maybe I should have brought that up"; "Didn't think of it"; or, "Maybe I should have done more with it". Her assessment indicators captured a comprehensive range of expressed morality, and her instincts, gut feelings, moral wisdom, and teaching experience guided her responses toward enhancing students' moral learning, growth, and development. Terry worries, nonetheless, that this inherent subjectivity would be difficult to justify, should she be asked to formally report on her students' moral progress to administrators, other teachers, students, and parents. Consequently, she maintains an interest in developing a protocol for the school's Character Development Program, which combines elements of her own approach with an objective measurement scale related to the school's values.

CONCLUSION

With four direct pedagogical practices—virtues instruction, discussions, discipline, and service activities—Terry conveyed morality to her students, as both knowledge content and abilities content. Delineated in Chapter Two, knowledge content supports the students' understanding of good and bad, and right and wrong, as applied to behaviour in social interaction. Virtues instruction was predominantly concerned with the transmission of this type of content. Abilities content supports the development of cognitive and emotional skills as applied to the processing of morally salient issues. Discussions were predominantly concerned with nurturing abilities content. Discipline methods and service activities each seemed to convey both knowledge and abilities by encouraging students to cultivate behaviours associated with integrity, responsibility, respect, consideration, empathy, and courage, among other moral values, and to identify choices, make decisions, solve problems, and resolve conflicts. Actively and continuously assessing how well the students were acquiring both types of content enabled Terry to adjust these practices and maximize their educative potentials.

As the students acquired moral knowledge and developed moral ability over the course of the school year, they were better able to foster and sustain mutually caring relationships with Terry and each other; to partner with Terry in nurturing a class community characterized by moral values of inclusiveness, respect, kindness, care, sensitivity, responsibility, and helpfulness, among others; and to ensure everyone's happiness at school, including Terry's. Thus, while teaching morality might seem to serve its own end, related to students' growth and development, it also facilitated Terry's ability to teach morally. Teaching morally and teaching morality, in fact, are reciprocally supportive. Teaching morally provides a fertile environment for teaching morality and indirectly, at minimum, imparts morality to students, as detailed in Chapter Four. Teaching morality provides students with the knowledge and skill, possibly also the motivation, to support teachers' efforts in teaching morally. To conceptualize the complexity and breadth of what this implies, the final chapter returns to the concept of moral agency.

REFERENCES

American Girl Library Series. (2003). *A smart girl's guide to friendship troubles: Dealing with fights, being left out & the whole popularity thing.* Middleton, WI: American Girl Publishing.

Balch, M. F., Saller, K., & Szolomicki, S. (1993). Values education in American public schools: Have we come full circle? *Association for Supervision and Curriculum Development, 94,* 34. Retrieved March 25, 2009, from http://www.eric.ed.gov/ERICDocs/data/ericdocs2sql/content_storage_01/0000019b/80/15/94/95.pdf.

Billig, S. H. (2000). The effects of service learning. *School Administrator, 57*(7), 14–18. Retrieved May 15, 2012, from http://proquest.com.

Billig, S. H. (2009). It's their serve. *Leadership for Student Activities, 37*(8), 8–13. Retrieved May 15, 2012, from http://proquest.com.

Campbell, E. (2003). *The ethical teacher.* Maidenhead: Open University Press McGraw-Hill.

Fenstermacher, G. D. (2001). On the concept of manner and its visibility in teaching practice. *Journal of curriculum studies, 33*(6), 639–653. doi: 10.1080/00220270110049886.

Fenstermacher, G. D., Osguthorpe, R. D., & Sanger, M. N. (2009, Summer). Teaching morally and teaching morality. *Teacher Education Quarterly, 36*(3), 7–19. Retrieved February 25, 2010, from http://vnweb.hwwilsonweb.com.

Furco, A., & Root, S. (2010, February). Research demonstrates the value of service learning. *The Phi Delta Kappan, 91*(5), 16–20. Retrieved May 22, 2012, from www.jstor.org.

Goodwill. (n.d.). About Us. Retrieved February 9, 2010, from www.goodwill.on.ca.

Hart, D., Matsuba, M. K., & Atkins, R. (2008). The moral and civic effects of learning to serve. In L. P. Nucci & D. Narváez (Eds.), *Handbook of moral and character education* (pp. 484–499). New York: Routledge.

Hartshorne, H., & May, M. A. (1930). A summary of the work of the character education inquiry. *Religious Education, 25*, 754–762. doi: 10.1080/0034408300250810.

Hildebrandt, C., & Zan, B. (2008). Constructivist approaches to moral education in early childhood. In L. P. Nucci & D. Narváez (Eds.), *Handbook of moral and character education* (pp. 352–369). New York: Routledge.

Hinck, S. S., & Brandell, M. L. (1999, October). Service learning: Facilitating academic learning and character development. *NASSP Bulletin.* Retrieved March 17, 2011, from http://proquest.umi.com.

Howard, R. W. (2005). Preparing moral educators in an era of standards-based reform. *Teacher Education Quarterly, 32*(4), 43–58. Retrieved February 26, 2010, from http://proquest.umi.com.

Howard, R. W., Berkowitz, M. W., & Schaeffer, E. F. (2004). Politics of character education. *Educational Policy, 18*(1), 188–215. doi: 10.1177/0895904803260031.

Jackson, P., Boostrom, R., & Hansen D. (1993). *The moral life of schools.* San Francisco: Jossey-Bass.

Lakin, R., & Mahoney, A. (2006). Empowering youth to change their world: Identifying key components of a community service program to promote positive development. *Journal of School Psychology, 44*, 513–531. doi: 10.1016/j.jsp.2006.06.001.

Landowne, Y. (2004). *Selavi: A Haitian story of hope.* El Paso, TX: Cinco Puntos Press.

Lickona, T. (1991). *Education for character: How our schools can teach respect and responsibility.* New York: Bantam Books.

Lickona, T. (2004). *Character matters: How to help our children develop good judgement, integrity, and other essential virtues.* New York: Touchstone.

Lowry, L. (1989). *Number the Stars.* New York: Yearling.

McClellan, B. E. (1999). *Moral education in America: Schools and the shaping of character from colonial times to the present.* New York: Teachers College Press.

Morgan, W., & Streb, M. (2001). Building citizenship: How student voice in service-learning develops civic values. *Social Science Quarterly, 82*(1), 154–169. doi: 10.11 11/0038–4941.00014.

Noddings, N. (2008). Caring and moral education. In L. P. Nucci & D. Narváez (Eds.), *Handbook of moral and character education* (pp. 161–174). New York: Routledge.

Nucci, L. P. (2009). *Nice is not enough: Facilitating moral development.* Upper Saddle River, NJ: Pearson Education.

Oser, F. (1994). Moral perspectives on teaching. In L. Darling-Hammond (Ed.), *Review of research in education* (pp. 57–127). Washington, DC: American Educational Research Association. Retrieved April 16, 2008, from http://www.jstor.org/stable/1167382.

Richardson, V., & Fenstermacher, G. D. (2000). *The Manner in Teaching Project.* Retrieved July 29, 2008, from www.personal.umich.edu/~gfenster.

Richardson, V., & Fenstermacher, G. D. (2001). Manner in teaching: The study in four parts. *The Journal of Curriculum Studies, 33*(6), 631–637.

Ryan, K., & Bohlin, K. E. (1999). *Building character in schools: Practical ways to bring moral instruction to life.* San Francisco: Jossey-Bass.

Schuitema, J., ten Dam, G., & Veugelers, W. (2008). Teaching strategies for moral education: A review. *Journal of Curriculum Studies, 40*(1), 69–89. doi: 10.1080/00220270701294210.

Schwartz, L. (1997). *Teaching Values—Reaching Kids*. Newton Centre, MA: Learning Works.

The Terry Fox Foundation. (n.d.). The Terry Fox National School Run Day. Retrieved September 27, 2009, from http://www.terryfox.org.

Watson, M. (2003). *Learning to trust: Transforming difficult elementary classrooms through developmental discipline*. San Francisco: Jossey-Bass.

Watson, M. (2008). Developmental discipline and moral education. In L. P. Nucci & D. Narváez (Eds.), *Handbook of moral and character education* (pp. 175–203). New York: Routledge.

6 Moral Agency

The Science and Art of Moral Education

During my year of fieldwork and the two years of writing that followed, I often asked myself, "What is holding all of this together?" By "this" I meant the class community; the caring relationships Terry fostered with and among students; the moral values infused into daily routines, activities, events, and social interactions; the moral sensibilities and awareness nurtured in students; the array of pedagogical strategies and practices for imparting morality; and the mostly hidden means by which Terry continually assessed students' moral learning, growth, and development. There were no files in the cabinet, no books on the shelves, and no resource materials in the cupboard that covered the range and complexity of this *course* on morality. There was no computer program or bookmarked website, no workshop package, no dedicated space in the classroom, and no time allotted in the timetable. The usual structures associated with school programs did not exist for the moral education Terry was providing. The answer to my question invariably rested on Terry and the morally laden vision she maintains. Objectives related to teaching morally and teaching morality, including nurturing a class community in which all students are happy and facilitating students' social-moral growth development, respectively, served as both beacons and yardsticks, enabling Terry to hold in her head, hands, and heart—as Sergiovanni (1992), Ryan and Bohlin (1999), and others might say—the entire moral landscape of this classroom and her students' experiences within it. This is a responsibility Terry recognizes and accepts:

> When you're looking at morals, it encompasses everything—how you do things *and* how you think about things. I wouldn't know how to compartmentalize and say, "I'm this kind of moral educator, therefore I can't do that activity because that's not who I am". The character education stuff should support care ethics and being in relationship with each other. And service, as well—caring more outside of you and the community. The developmental approach is how you develop as a person. Also your moral reasoning, so you can problem-solve and make decisions that support being in relationships. And development of your character—everybody's at different stages there, too. So that's why

with Connor and Kay, they understand more deeply. It means something more to them. But for others, it may not mean the same. So if you
are not a service-oriented kind of moral educator, then how are you
going to work with somebody who is showing more affinity towards
that end? You have to be all of that, and maybe some other things that
haven't been identified, to make sure you can support the full range of
development.

Being "all of that" and more not only involves attending to each of the
themes referenced by Terry in the passage above and delineated in previous
chapters, but also involves integrating these themes in practice. Previous
empirical work, particularly the qualitative studies outlined in Chapter Two,
identified many similar themes among the practices of several different
teachers. This portrait also reveals how these themes might be integrated
within the practice of a single teacher. I have adopted and adapted the term
moral agency by way of conceptualizing the breadth and complexity that
this entails. This final chapter expands the definition of moral agency presented in Chapter One and speculates on implications for teaching practice,
teacher education, and academic research to suggest how we might move
forward in envisioning and enacting classroom-based moral education.

TERRY'S PRACTICES AS SCIENCE AND ART

Creating this portrait, I was challenged to represent the holistic nature of
Terry's moral agency while also presenting an analytic discussion. Analysis
required that significant themes be abstracted. Doing so, however, risked
inauthenticity because Terry does not think of or enact her practices in this
way, although she often could appreciate the particular themes when questioned, articulating versions of "I never thought that that's what I did till
you mentioned it. But that's definitely what I do". Consequently, I struggled
against intangible forces, variously pulling together different aspects of the
themes and hindering the establishment of clean categories. Webs material-
ized, cross-connecting practices and strategies, philosophical orientations,
moral values, and domains of practice.

As might have been evident in Chapters Four and Five, three webs integrate Terry's practices and strategies. The first relates to the objectives they
serve. For example, the personal interactions Terry initiates with students
by way of fostering relationships with them are facilitated by expressions
of fairness, respect, kindness, and honesty in her conduct and behaviours.
Expressing these and other moral values also enables Terry to preserve students' dignity during discipline. Terry's individual relationships with each
student support a class community by setting a standard for caring relationships among them. Class community provides a favourable context in which
respectful class-wide discussions might take place. Discipline and discussions

often convey virtues messages. Lastly, discussions on morally salient topics can be a source of service activities.

In the second web, practices and strategies are associated through dichotomous descriptors, particularly indirect and direct, and formal and informal. Indirect practices and strategies refer to moral education that is implicitly mediated through, for example, teachers' conduct and behaviours, the classroom environment, relationships, and seating arrangements, as well as discussions on issues that are not moral in nature. Developmental discipline, discussions on moral issues, virtues instruction, service learning activities, and collaborative classroom processes involve the overt and explicit expression of moral values and, thus, are considered to be direct. Students might not be aware of moral values that are conveyed indirectly but are indeed aware when conveyed directly. Formal practices and strategies are preplanned or established features of curriculum and classroom life, such as virtues *lessons*, greeting students at the door, Terry's welcome package, adult-initiated service activities, and the process of rotating seating arrangements. Informal practices and strategies arise spontaneously. These include, for example, virtues *messages*, showcasing, disciplining, discussions, and acts of consideration and kindness. Connections are evident. Indirect practices and strategies might be enacted formally, as with the system for rotating seating arrangements, or informally, as with modelling moral conduct. Direct practices and strategies might be enacted formally, as with virtues lessons, or informally, as with virtues messages. Some scholars apply these and other dichotomous descriptors to distinguish and advocate for particular orientations to moral education (Leming, 2008; Sanger & Osguthorpe, 2005). For example, character education might be promoted as direct and formal and Kohlberg's Just Communities as indirect, with both formal and informal qualities.

Finally, Terry's practices and pedagogical strategies indicate a web among philosophically distinct orientations for moral education. Several practices and strategies are promoted by more than one orientation and, thus, serve more than one educative outcome. For example, modelling moral behaviour is recommended for character education (Lickona, 1991; Ryan & Bohlin, 1999; Wynne & Ryan, 1997) and care ethics (Noddings, 2002). Thus, in demonstrating good and right behaviours for students to emulate and habituate, Terry supports the development of her students' character dispositions and facilitates the reciprocal nature of the caring relationship they develop with her. Utilizing academic curricula to impart moral messages is recommended for character education (Lickona, 1991, 2004; Lickona & Davidson, 2005; Ryan & Bohlin, 1999; Wynne & Ryan, 1997), care ethics (Noddings, 2002), and cognitive development (Nucci, 2009). Accordingly, Terry illustrates how the language arts curriculum helps her to nurture both character development and cognitive development among her students:

> If you could look at a whole bunch of stories and focus on the moral lessons and values in them, then [the students] become more aware of

morals and values in whatever they're reading and writing. It comes out in their work later on. I can ask them to discuss a certain value or why a character did something that was right or something that was wrong. I would be able to see, through explicit examples, how their moral thinking was progressing.

Fostering relationships with students, attending to classroom environments, providing opportunities for practicing desired moral behaviours, and encouraging dialogue and discussion are recommended for character education (Lickona, 1991, 2004; Lickona & Davidson, 2005; Ryan & Bohlin, 1999; Wynne & Ryan, 1997), care ethics (Noddings, 2002, 2008), and cognitive development (Nucci, 2009; Watson, 2003). In regard to discussions, for example, Terry believes that if students are not respectful in how they participate, their peers may be reluctant to assume responsibility for sharing thoughts and ideas. If thoughts and ideas are not freely shared, then students are not exposed to alternative ways of thinking, nor do they have their own thoughts challenged. Consequently, Terry communicates and upholds behavioural expectations that students will listen to each other respectfully and assume responsibility for sharing. This supports their character development, particularly in regard to habituating expressions of respect and responsibility. When the discussion involves a morally salient issue, Terry encourages students to probe each other's sensitivities, understandings, choices, and decisions, according to what is right and good. This reinforces moral reasoning. Discipline practices are similarly enacted in regard to respectful listening and responsible sharing, as well as identifying choices that are right and good. In addition, Terry considers her personal relationship with each student to be essential in the context of discipline so that students might receive correction from her without embarrassment or loss of dignity.

This notion of compatibility among different philosophical orientations has been conceptually explored. Noddings (2008) notes, "There are today several influential approaches to moral education . . . Except at the extremes, they are not in irresolvable conflict" (p. 168). Accordingly, Noddings (2002) attempted to reconcile care ethics and character education. Similarly, but perhaps more inadvertently, character educators make recommendations that coincide with care ethics, such as creating caring communities and fostering caring relationships. For example, Lickona (2004) sounds much like Noddings in asserting, "Schools must foster that same spirit of inclusiveness and the actual experience of caring community through day-to-day relationships" (p. 20); and "Good teachers build the relationship in both directions; they and their students learn about each other" (p. 115). Ryan and Bohlin (1999), also promoters of character education, posed the following series of practical questions, implying strategies that might be endorsed by care ethicists:

Does the teacher respect the students? Do the students respect one another? Are classroom rules and teacher expectations fair? Are they justly enforced? Does the teacher play favourites? Are ethical questions such as "What is the right thing to do?" part of the classroom dialogue? (pp. 144–145)

Nucci (2009) may endorse the last question, as well, in accordance with cognitive development theory.

There have also been attempts to reconcile theories of cognitive development and care ethics (Johnston, 2006; Noddings, 2008). With roots in psychology, both address innate developmental and social processes. Cognitive development focuses on autonomous reasoning, while care ethics focuses on affective relationships (Howard, 2005; Schuitema, ten Dam, & Veugelers, 2008). They are understood as complementary (Colnerud, 2006) or as "two moral predispositions that inhere in the structure of the human life cycle. Predispositions toward justice and toward care" (Gilligan, 1988, p. 4). Consequently, several recommendations made in a context of care ethics, including, for example, constructivist pedagogies and collaborative problem solving, are similarly found in Kohlberg's Just Communities, for promoting cognitive development.

Finally, Lockwood (2009) proposed a blend, of sorts, between character education and cognitive development, called developmental character education. This approach entails adding cognitive and affective attributes, associated with cognitive development, to character education, such that age-appropriateness, sound moral reasoning, perspective taking, and empathy are considered when teaching expressions of virtues and moral values. Coincidentally, Sean articulated a similar desire to delineate a developmental progression for each of Middlevale's values: "what it looks like to be courageous, for example, at 8 years, at 11 years, at 14 years". Should this be achieved, it would consolidate the school's character program with its developmental approach to discipline.

Terry's practices provide empirical evidence that aspects of these three philosophical orientations might be compatibly integrated in the classroom. Yet, Terry does not conceptualize the moral education she provides for students according to a particular theoretical position, ethical perspective, or set of assumptions within which these orientations are encased. Thus, the three webs among Terry's practices and strategies seem to suggest that reconciliation is realized in practice, not philosophy or theory, and that conceptual contradictions among the orientations might be understood as serving different outcomes and needs in a comprehensive approach to moral education, rather than as alternatives from which teachers must choose only one.

Associations among the many moral values that Terry knowingly and intentionally operationalizes in the classroom create another web. Wynne and Ryan (1997) recommend that teachers focus on no more than six to eight virtues in order to avoid overlaps. In Chapters Four and Five, however, I note fairness, equity, equality, justice, respect, kindness, honesty, helpfulness, cooperation, care, compassion, responsibility, courage, integrity, empathy, sensitivity, acceptance, tolerance, thoughtfulness, consideration, inclusiveness, trust, self-respect, self-regulation, and openness. Overlaps are evident in how these moral values were expressed, particularly in regard to fairness, equity, and equality; kindness, care, compassion, sensitivity, and empathy; honesty, integrity, and openness; helpfulness and cooperation; acceptance, inclusiveness, and tolerance; thoughtfulness and consideration; courage and self-respect; self-regulation and responsibility. In fact, rarely, if ever, were moral values expressed in isolation of others, and the association among moral values often legitimized moral salience and demonstrated varied and nuanced manifestations in social interaction. This web suggests, therefore, that redundancy is an asset for moral education and that attempts to cap the number of moral values intentionally conveyed to students is unrealistic and inauthentic.

Finally, a web is evident among the four domains of teaching practice—teachers' vision, content to be taught and learned, teaching strategies, and methods of assessment. Terry's vision of classroom life and students' growth, development, and well-being influences the moral knowledge and abilities content she conveys to students. Content is conveyed using strategies for teaching morally and teaching morality. Terry assesses the effectiveness of her strategies, what content has been learned, and to what degree against her vision. Assessment informs the continued use of strategies and content still to be conveyed. Thus, employing strategies to impart morality, without a vision of the desired end, may limit one's ability to assess the usefulness and effectiveness of these strategies, causing the teacher, and possibly the students, to feel confused. Envisioning a desired end, without utilizing suitable strategies, may limit one's ability to close the gap between vision and reality, causing the teacher to feel frustrated. Utilizing particular strategies, without knowledge of the moral content being conveyed, may limit the ability of those strategies to achieve professed outcomes, causing the teacher to feel unconfident. This web suggests that educators carefully consider all four domains in order to impart moral education with clarity, satisfaction, and confidence.

The last web, more so than the others, is dynamic in nature, with cyclical and forward motions reminiscent of a spiral. The cyclical motion is derived from a feedback loop between vision and assessment, which is galvanized by continual adjustments to content and strategies. The forward motion is derived from Terry's vision. Terry believes humans to be in a constant state of development, with enormous potential for goodness. Accordingly, she does not sustain a maximum standard that her students might attain. Regardless of how much the students progress, individually and collectively, Terry modifies, increases, or intensifies her expectations, encouraging the students to be kinder and more considerate when interacting with others; to think more deeply, be more observant, and incorporate more contextual information when making decisions and acting on those decisions; and to take on more responsibility for themselves, each other, and the larger communities with which they are affiliated. Terry illustrates her advancing expectations with the following examples:

> First term felt more like a term that was really training them in being aware of others, being more cooperative, being more respectful and listening to others. Because at the beginning of the year, I'm trying to develop this sense of community. . . . With second term, the expectation is slowly putting more and more responsibility on them so that they need to think about where they're at and what they need to begin to be more self-aware. And I've talked about being more aware of others and making sure that others are happy or not unhappy—"You do something about it". Also working towards becoming more self-aware so that they can think things through and reflect more and more, rather than me always telling them what to do and how to do things. Giving them more independence in that decision-making would give them more confidence in the end. I don't want to have to make all the decisions myself. . . . Third term they have to do it by themselves now. I'm out of it—"Show me you can do it".

Consequently, formal lessons on compassion, cooperation, and respect occurred during the first term as a means of quickly and concisely communicating Terry's expectations for how she and the students were to interact with each other. In the second term, Terry built on the foundation of a class community by more specifically encouraging the students to partner with her in problem-solving, decision-making, and conflict resolution. The current events unit was also intentionally situated in this term to ensure the classroom environment was psychologically safe for dialogue and debate on a variety of issues. Terry's third-term expectation obligated students to assume more responsibility, not only for themselves but also for each other and the class as a whole. This was supported by a direct lesson on responsibility; increased levels of empowerment and trust; additional tasks; and a new related poster. These expectations accumulated over the school year but were never fully achieved. Yet, progress was regularly recognized, with variations of "They're coming along".

These webs are as integral to this portrait as the individual themes they connect. Not only does Terry attend to each of the themes previously discussed, she navigates the spaces between them in complex ways that remain somewhat elusive. In any given moment, Terry employs combinations of practices and pedagogical strategies and conveys a variety of moral values, not necessarily resting upon any one in particular but nimbly travelling the webs by which they are connected. Herein lies the science and art of teaching practice as applied to moral education. Dewey (1929) claimed that education is both a science and an art. He defined science as methodology, rather than subject-specific knowledge: "The existence of systematic methods of inquiry, which, when they are brought to bear on a range of facts, enable us to understand them better and to control them more intelligently, less haphazardly and with less routine" (pp. 8–9). Terry has benefitted from the accumulated theoretical, philosophical, practical, and empirical scholarship on moral education. This is noted in the reflective way in which she discusses and employs particular practices and strategies reported in education literature. As an art, Dewey likens education to "a kind of social engineering" (p. 39). Accordingly, one "make[s] new integrations of scientific material and turn[s] it to new and previously unfamiliar and unforeseen uses" (pp. 13–14). Terry's rejection of a prepackaged character program for the school, in favour of a locally developed version, her less than enthusiastic approach to character worksheets, and her integration of practices, strategies, and orientations demonstrates a determination to not rely on or be confined to the science of moral education while still being informed by it.

There is no contradiction between education as science and education as art, in either Dewey's characterization or Terry's practice. One supports the other. Practitioners' observations, reflections, inquiry, instincts, judgments, and speculations inform personal and general practice, giving rise to a "steady and cumulative growth of intelligent, communicable insight and power of direction" (Dewey, 1929, p. 10). Accordingly, Terry's practices, as portrayed in this portrait, might be considered a contribution to the science of moral education. Further, "the existence of science gives common efficacy to the experiences of the genius; it makes it possible for the results of special power to become part of the working equipment of other inquirers, instead of perishing as they arose" (Dewey, 1929, p. 11). Terry acknowledged that her research on moral education did provide structure and a framework for the moral education she was already imparting in the classroom. Thus, Terry is both scientist and artist, as she integrates education literature with personal and folk wisdom, methodology with improvisation, and strategy with instinct.

MORAL EDUCATION AS MORAL AGENCY

In Chapter One, I repositioned Campbell's (2003) two-pronged definition of moral agency to focus on the teacher as a moral educator who upholds as a primary concern, the moral learning, growth, and development of students. Inevitably, this teacher is also a moral person. In Chapters Four and Five, I applied Fenstermacher, Osguthorpe, and Sanger's (2009) notions of teaching morally and teaching morality to discuss Terry's moral education practices and strategies. Uniting these two frameworks proposes that a moral agent teacher is a moral educator who knowingly and intentionally imparts morality to students, as well as a moral person who teaches morally or in morally justifiable ways. As such, the moral agent teacher attends to the moral dimensions of teaching, learning, and classroom life and to the teaching dimensions of morality. The former phrase is familiar. The latter phrase acknowledges distinct qualities related to teaching morality that this portrait reveals—for example, the often spontaneous nature of moral messages and the ability of teachers to seize morally salient issues that arise in classroom circumstances.

Further, I consider ethical knowledge to be the knowledge base of moral agency. Defined by Campbell (2003) as the application of one's personal sense of morality to professional practice, ethical knowledge brings to consciousness the conceptual and practical connections between personal interpretations of moral principles and virtues and professional obligations, duties, roles, and responsibilities; it also illuminates the moral complexities, conflicts, and nuances of classroom life, teaching practice, and learning goals. As such, ethical knowledge enhances one's ability to attend to the moral dimensions of teaching, learning, and classroom life. In addition, Terry's ethical knowledge informs her vision for classroom life and students' growth, development, and well-being; enables her to regularly seize the more embedded opportunities for imparting morality to students as moral education; and facilitates her ability to assess the moral development of her students and readjust her practices and goals as necessary. Therefore, ethical knowledge is also understood as enhancing one's ability to teach morality.

Moral agency, defined as teaching morally and teaching morality, offers a comprehensive conceptualization of moral education in the classroom. It is independent of any particular ethical, theoretical, and philosophical orientation, yet able to embrace the assumptions, beliefs, and pedagogies of more than one. In addition, it honours and respects teachers' folk wisdom, professional experiences, personal instincts, gut feelings, and moral understandings. It acknowledges the tangible and delineable themes, as well as the intangible webs and spaces between them. It endorses prevailing as well as personally developed teaching practices. Finally, it is consistent with results of other empirical investigations, including The Moral Life of Schools Project (Jackson, Boostrom, & Hansen, 1993), The Manner in Teaching Project (Fenstermacher, 2001; Richardson & Fenstermacher, 2000, 2001), and the Moral and Ethical Bases of Teachers' Interactions with Students study (Campbell, 2003, 2004; Campbell & Thiessen, 2000, 2001). The Moral Life of Schools Project, in particular, also recognized a dual nature to moral education, categorizing teachers' moral influence as moral *practice* and moral *instruction*. This coincides with the moral person and teaching morally, and the moral educator and teaching morality, respectively. Moral education as moral agency might, therefore, help to address Sanger and Osguthorpe's (2005) concern that "the conceptual geography of moral education scholarship and research makes it difficult to synthesize current understanding in a way that allows us to effectively represent, critically analyze, refine and further that understanding" (p. 58).

MOVING FORWARD

Previous empirical work enabled a composite image of moral agency. This portrait demonstrates that such a teacher does, in fact, exist and provides an exemplar to which other teachers might reasonably aspire. Previous work revealed multiple methods, mechanisms, practices, approaches, orientations, programs, and frameworks that teachers utilize to impart moral messages and lessons. This portrait offers an opportunity to understand how combinations of these may operate harmoniously and compatibly within one classroom and be dynamically managed by a single teacher. Finally, previous work focused on the content of moral messages and lessons, and on strategies for imparting morality. This portrait also adds insights related to teachers' moral vision and assessment methods. These new contributions hold implications for teaching practice, teacher education, and academic research.

Teaching Practice

Leming (2008) warns, "Unless research addresses practice in a way that is perceived by teachers as clear, salient, and utilitarian, it will likely remain irrelevant to classroom practice" (p. 135). Accordingly, I gave Terry a strong voice in representing her moral agency and have provided copious and contextualized descriptive detail, hoping that educators will find resonance with their own practices and situations and be able to make decisions on transferability, at their discretion. As Terry notes, "There are certain things which are general enough to be able to transfer". A single-case study, however, does not enable me, as researcher, to generalize beyond this teacher and her teaching context. Thus, I do not provide heuristic devices or make prescriptive recommendations, and this portrait should not be understood as a blueprint or checklist for moral education.

However, it is noteworthy that this portrait coincides with what many teachers already do in the classroom. Without requiring an entire revision, therefore, the portrait has the potential to renew classroom-based moral education in three ways. First, for teachers who presently give minimal priority to moral education, any aspect of this portrait can provide a focus, structure, or rallying point around which a comprehensive approach might develop. This was the case with Middlevale's Character Development Program, which evolved from a reassessment of school discipline. Similarly, the earliest expressions of Terry's moral agency were the community service activities she initiated with her classes, as a novice teacher. Fostering relationships with students, nurturing community, intentionally modelling moral values, virtues instruction, and class-wide discussions might also be starting points, in accordance with a teacher's ethical orientation, teaching style, school context, or area of interest.

Second, teachers who have integrated moral education into their practice might use this portrait to expand their repertoire of strategies. For example, those who deliver formal virtues lessons may be inspired to more intentionally complement these with spontaneous virtues messages, or the inverse. Teachers who focus on relationships to foster a class community might also see the value of empowering students to engage in collaborative problem-solving, decision-making, and peer mediation. Teachers who mediate class discussions on moral issues might seize the opportunity to reinforce pro-social values of listening and candidness and moral values of respect and responsibility. Terry's experiences demonstrate that moral agency is developed over time, as practices for teaching morally and teaching morality are acquired and accumulate.

Finally, regardless of whether teachers are at the earliest or more advanced stages as moral educators, pondering the analytic discussions and anecdotal accounts of this portrait can enhance their ethical knowledge. Analytic discussions deconstruct the moral intent of a variety of practices and strategies. Anecdotes reveal the nuances and complexities by which they are enacted. Further, anecdotes were situated to highlight particular themes but invariably involve more than a single theme. Thus, teachers might explore additional aspects of moral agency in each and be motivated to similarly consider their own practices through a moral lens. While I cannot conclude that in doing so they will become moral agents, Campbell's (2003) empirical work indicates that moral agent teachers have high levels of ethical knowledge, and my work with Terry indicates that reflecting on practice through a moral lens enhances one's ethical knowledge.

Teacher Education

Pre-service teacher educators are implored to cultivate among their students, suitable sensibilities, awareness, and practices related to teaching morally and teaching morality (Sanger & Osguthorpe, 2013). Some scholars suggest building on the ideals that attracted teacher candidates to the profession. Ryan and Bohlin (1999) recognize that "the majority of teachers enter the profession with the goal of devoting their life to the betterment of the young . . . helping children become better human beings, better people" (p. 152). Terry agrees: "Teachers, in general, are that kind. They do see something rewarding in what they do because of good reasons. It's there. They want to do the service, and they want to help in some way". Accordingly, O'Sullivan (2005) proposed:

> Teacher education programs must nurture the deeper dreams future teachers bring to their course work. These dreams involve helping children become better people, not just smarter people. Yet these dreams can only be carried out by teachers who are themselves becoming better as well as smarter. There is an old saying in teaching that if you don't feed the teachers, they will eat the children. (p. 8)

Developing the teacher as a moral person and a moral educator was similarly recommended by Schwartz (2008), who notes, "Tomorrow's teachers need to do two things: they need to develop their own moral and ethical character so they can lead by example and they need to learn the *pedagogy* [italics in original] of moral and character education" (p. 597). Moral agency, as advanced in this portrait, enables me to speculate on how such development might be facilitated.

Terry acknowledges that "teachers could be brought along as moral educators, or moral agents, or character educators". Yet, in reflecting on her own experiences, she wonders about the timing of such expectations:

> As a newer beginning teacher, I was more concerned about doing my job right. I was more concerned about making sure I did everything I was supposed to do. I was more worried about the lesson plan and what pages to get through and how to teach it, how I'm going to teach this lesson. And until you feel comfortable with everything that's going on around you, whether it's forms that you have to fill out, school meetings, all of the day-to-day scheduling and also curriculum, until you're comfortable with that it's hard to talk to somebody about the other things that you now need to look at.

Thus, despite entering the profession with ideals related to helping and caring for children, in the early years Terry inevitably focused on delivering curricula, managing the class, and participating as a member of the school faculty. Efficiency, safety, and effectiveness were the values that guided her practice. Only after she had matured into the role and responsibilities was Terry able to return to her ideals with a heightened sense of moral purpose. As a result, Terry questions the practicality and reasonableness of trying to nurture the full expression of moral agency during pre-service.

It cannot, nonetheless, be assumed that all teachers will revisit and reengage rudimentary ideals, as Terry did, once these ideals have been overshadowed or displaced by the technical and mechanical aspects of the job. Terry and I discussed, therefore, the possibility of a two-tiered approach to nurturing moral agency, which takes place over several years. The first tier occurs during the pre-service program, where the ideals of teacher candidates might be validated, explored, and shaped into concrete moral visions (Sanger & Osguthorpe, 2011). Terry suggests that teacher educators encourage their students to reflect, as follows:

> What do you care most about? Why do you do this? What would be the driving force in you as a teacher? Why do you want to teach children? What does it mean for you to be here? What's the purpose of you being here? What's the purpose of what you want to do? That is the first thing maybe. Because if you don't know why you do it, then there's no driving organizational framework from which everything stems. That's how

I approach everything myself. . . . To pick and choose the kinds of activities and the way that I approach things has to come from something inside. Or it's not going to be true. I reflect on myself first because that's what makes the teacher a moral agent.

Additionally, teacher candidates can be exposed to the moral dimensions of professional practice as they become relevant in their program. This includes, for example, the role of fairness and honesty in assessment; trust and respect in classroom management; justice in discipline; equity and responsibility in the delivery of curricula; and courage and empathy in communication with parents. Thus, ethical knowledge and pedagogical knowledge might be simultaneously acquired. Several teacher educators present their programs and pedagogical approaches for doing so in *The Moral Work of Teaching and Teacher Education* (Sanger & Osguthorpe, 2013). As ideals, vision, and ethical knowledge underpin both teaching morally and teaching morality, new teachers might be motivated and encouraged to explore a variety of related practices.

The second tier occurs once teachers have become comfortable and competent with the mechanical and technical elements of their practice and have accumulated classroom experiences. Accordingly, it takes place in-service, as professional development. Teacher mentors might work closely with less experienced teachers, in the context of the trainees' classrooms and existing practices, to enhance the sophistication of their ethical knowledge and to expand their repertoire of strategies for teaching morally and teaching morality. Terry compares such mentoring to the experience of having me in her classroom:

> With respect to the moral education, I don't think about it as much anymore, since the character education research. And I had to really think about it this year, just to make sure that I was still on top of it. So I think that was also helpful. You made me bring it back into focus, I guess. It's always something that I do, but I never really think about it as much as I did again this year because you asked me to, which is good. I'm more conscious of it with you.

Notwithstanding Terry's endorsement, the feasibility and efficacy of this approach to nurturing moral agent teachers requires further investigation.

Academic Research

To complement this portrait of moral agency, I propose that academic research continue in three directions. In regard to the first, Terry is not the only teacher who might be considered an exemplar of moral agency, and I am not the only researcher who might investigate this phenomenon. Conducting similar micro-ethnographic studies with different researchers and teachers in a variety of teaching contexts will serve to strengthen this conceptualization of moral agency by deepening, broadening, and adding nuance to its themes. For example, a male teacher might reveal gender differences related to vision, ethical orientation, and assessment. Alternative school contexts, such as residential or publicly funded, and alternative classroom contexts, such as special education or single-gender, might reveal challenges and limitations necessitating unanticipated strategies. Additional opportunities, strategies, and challenges might also be revealed in schools that do not maintain a moral education infrastructure or explicitly encourage moral education in classrooms. Lastly, a researcher with a critical or social justice perspective might interpret the teacher's approach to service activities differently, and a male researcher might differently characterize the class community. These further studies also have the potential to increase our understanding of the more veiled aspects of moral agency, such as how one navigates among the various practices and the role of vision in defining a moral education agenda.

The second direction for continued research involves a collaborative and long-term relationship between teachers and researchers, such that ideas generated from discussion and observation might be simultaneously operationalized and studied. For example, following our discussion on assessment, Terry and I might have developed and implemented an objective measure for her students' moral progress or a school-wide measure to complement the character program. Similarly, following the discussion on teacher education we might have developed and implemented a mentoring protocol to nurture moral agency among the faculty. The feasibility and efficacy of both initiatives could then have been studied, in action, along with any refinements or recreations. Such studies provide results that are accessible to practitioners and directly applicable to practice.

Finally, investigations that ascertain the effectiveness of teachers' efforts to impart morality are indicated. The literature is saturated with recommendations regarding how and why to deliver moral education. Yet, many professed short-term and long-term outcomes on students' behaviours, conduct, relationships, choices, attitudes, motivations, understandings, and cognitive processing abilities are hypothetical and speculative. Further, it is unclear what students understand about their teachers' ethical or unethical conduct, behaviours, and practices; if students recognize the efforts of educators to provide a moral education alongside the academic education; and how applicable students believe moral messages and lessons are to their out of school lives, both in the present and in the future. Students' actual learning, development, experience, and perspective are underrepresented in moral education literature. Adding related insights would substantiate and guide educators' continued efforts toward moral education to ensure they are well placed and meaningful.

These proposals for teaching practice, teacher education, and academic research imply that moving forward with moral education as moral agency involves the participation of practitioners, scholars, and students, both within the scope of their own roles, responsibilities, and experiences as well as in collaboration with each other. The participation of practitioners in research activities, for example, can help to ensure that theory both reflects practice and is applicable to practice, so that the gap between theory and practice (Leming, 2008) is lessened. The participation of scholars can help to ensure that insights from practice are integrated, deeply conceptualized, and widely disseminated, such that teacher educators and mentors can effectively nurture moral agency among candidate and novice teachers. The participation of students can help to ensure that the efforts put forth by educators resonate with what students deem to be important and necessary for their moral learning, growth, and development. Encouraged by Burkhardt and Schoenfeld's (in Leming, 2008) assertion that research should be less concerned with developing generalizable views about teaching practices and more concerned with generating high quality discussions, I present this portrait of moral agency, not as a definitive product but as a work in progress. I beseech practitioners, scholars, and students to contribute to it by way of advancing classroom-based moral education that is comprehensive, authentic, meaningful, and effective.

REFERENCES

Campbell, E. (2003). *The ethical teacher*. Maidenhead: Open University Press McGraw-Hill.

Campbell, E. (2004). Ethical bases of moral agency in teaching. *Teachers and Teaching: Theory and Practice, 10*(4), 409–428.

Campbell, E., & Thiessen, D. (2000, April). *Moral and ethical exchanges in classrooms: Preliminary findings*. Paper presented at the Annual Meeting of the American Educational Research Association, New Orleans, LA.

Campbell, E., & Thiessen, D. (2001, April). *Perspectives on the ethical bases of moral agency in teaching*. Paper presented at the Annual Meeting of the American Educational Research Association, Seattle, WA.

Colnerud, G. (2006). Teacher ethics as a research problem: Syntheses achieved and new issues. *Teachers and Teaching: Theory and Practice, 12*(3), 365–385.

Dewey, J. (1929). *The sources of a science of education*. New York: Horace Liveright. Retrieved July 18, 2014, from https://archive.org/details/sourcesofascienc 009452mbp.

Fenstermacher, G. D. (2001). On the concept of manner and its visibility in teaching practice. *Journal of curriculum studies, 33*(6), 639–653. doi: 10.1080/00220 270110049886.

Fenstermacher, G. D., Osguthorpe, R. D., & Sanger, M. N. (2009, Summer). Teaching morally and teaching morality. *Teacher Education Quarterly, 36*(3), 7–19. Retrieved February 25, 2010, from http://vnweb.hwwilsonweb.com.

Gilligan, C. (1988). Remapping the moral domain: New images of self in relationship. In C. Gilligan, J. V. Ward, & J. M. Taylor (Eds.), *Mapping the moral domain* (pp. 3–19). Cambridge, MA: Harvard University Press.

Howard, R. W. (2005). Preparing moral educators in an era of standards-based reform. *Teacher Education Quarterly, 32*(4), 43–58. Retrieved February 26, 2010, from http://proquest.umi.com.

Jackson, P., Boostrom, R., & Hansen, D. (1993). *The moral life of schools*. San Francisco: Jossey-Bass.

Johnston, D. K. (2006). *Education for a caring society: Classroom relationships and moral action.* New York: Teachers College Press.

Leming, J. S. (2008). Research and practice in moral and character education: Loosely coupled phenomena. In L. P. Nucci & D. Narváez (Eds.), *Handbook of moral and character education* (pp. 134–157). New York: Routledge.

Lickona, T. (1991). *Education for character: How our schools can teach respect and responsibility.* New York: Bantam Books.

Lickona, T. (2004). *Character matters: How to help our children develop good judgement, integrity, and other essential virtues.* New York: Touchstone.

Lickona, T., & Davidson, M. (2005). *Smart & Good High Schools: Integrating excellence and ethics for success in school, work, and beyond.* Cortland, NY, Washington, DC: Center for the 4th and 5th Rs/Character Education Partnership. Retrieved January 3, 2009, from http://www.cortland.edu/character/.

Lockwood, A. L. (2009). *The case for character education: A developmental approach.* New York: Teachers College Press.

Noddings, N. (2002). *Educating moral people: A caring alternative to character education.* New York: Teachers College Press.

Noddings, N. (2008). Caring and moral education. In L. P. Nucci & D. Narváez (Eds.), *Handbook of moral and character education* (pp. 161–174). New York: Routledge.

Nucci, L. P. (2009). *Nice is not enough: Facilitating moral development.* Upper Saddle River, NJ: Pearson Education.

O'Sullivan, S. (2005). The soul of teaching: Educating teachers of character. *Action in Teacher Education, 26*(4), 3–9. Retrieved December 10, 2009, from http://vnweb.hwwilsonweb.com.

Richardson, V., & Fenstermacher, G. D. (2000). *The manner in teaching project.* Retrieved July 29, 2008, from www.personal.umich.edu/~gfenster.

Richardson, V., & Fenstermacher, G. D. (2001). Manner in teaching: The study in four parts. *The Journal of Curriculum Studies, 33*(6), 631–637.

Ryan, K., & Bohlin, K. E. (1999). *Building character in schools: Practical ways to bring moral instruction to life.* San Francisco: Jossey-Bass.

Sanger, M., & Osguthorpe, R. (2005). Making sense of approaches to moral education. *Journal of Moral Education, 34*(1), 57–71. doi: 10.1080/03057240500049323.

Sanger, M. N., & Osguthorpe, R. D. (2011). Teacher education, preservice teacher beliefs, and the moral work of teaching. *Teaching and Teacher Education, 27*, 569–578.

Sanger, M. N., & Osguthorpe, R. D. (Eds.). (2013). *The moral work of teaching and teacher education. Preparing and supporting practitioners.* New York: Teachers College Press.

Schuitema, J., ten Dam, G., & Veugelers, W. (2008). Teaching strategies for moral education: A review. *Journal of Curriculum Studies, 40*(1), 69–89. doi: 10.1080/00220270701294210.

Schwartz, M. J. (2008). Teacher education for moral and character education. In L. P. Nucci & D. Narváez (Eds.), *Handbook of moral and character education* (pp. 583–600). New York: Routledge.

Sergiovanni, T. J. (1992). *Moral leadership: Getting to the heart of school improvement.* San Francisco: Jossey-Bass.

Watson, M. (2003). *Learning to trust: Transforming difficult elementary classrooms through developmental discipline.* San Francisco: Jossey-Bass.

Wynne, E. A., & Ryan, K. (1997). *Reclaiming our schools: Teaching character, academics, and discipline.* Upper Saddle River, NJ: Prentice-Hall.

Index